Spotter's Guide to
WILD PLANTS

Edited by
Sue Jacquemier and
Patricia Monahan

Designed by
Valerie Sargent and
Cloud Nine

Consultants
Richard Clarke
Esmond Harris
Chris Humphries
Peter Smith
Ian Tittley

Illustrated by
John Barber, Joyce Bee,
Mark Burgess, Hilary Burn,
Andy Martin, Dee McLean,
Annabel Milne & Peter Stebbing,
Julie Piper, Cynthia Pow

First published in 1981 by
Usborne Publishing Limited,
20 Garrick Street, London WC2

© 1981 by Usborne Publishing Limited

Printed by
Mateu Cromo Artes Gráficas, S.A.
Madrid, Spain

Spotter's Guide to
WILD PLANTS
Trees, Flowers, Fungi, Seaweeds, Ferns, Grasses, Mosses, Lichens and Liverworts

Contents

How to use this book

This book is about some of the plants which grow wild in Britain and other parts of Europe and it contains descriptions of more than 380 different species of wild plants. The ones that have been illustrated were chosen because they are particularly common, or because they are typical of a particular group. Some are included because they are rare, and some simply because they are beautiful.

Many interesting and useful wild plants are now rare, mainly because man has destroyed their natural habitats. The more we know and appreciate plants, the more we will be able to conserve.

How the book is organized

On the opposite page there is a list of the subjects dealt with in each Chapter. All the Chapters start with an introduction. This will give you some general information about that particular group of plants, their characteristic features, where they live and how they live. The introduction is followed by several pages which illustrate many of the common species of that group, as well as some less common ones.

The illustrations

Each illustration is designed to give you as much information as possible. Where we think it will be useful, we have provided a more detailed drawing of a particularly good identifying feature. Beside each illustration there is a brief description of the plant and its habitat. We have also given you a measurement – usually the height from the ground. This is an important piece of information, since the plants are not always drawn to scale.

Next to each description you will find a small circle. You could use this for ticking off each species as you find it.

On pages 178-9 are some hints on how to study and collect wild plants without endangering their survival.

The glossary provides definitions of some specialist terms.

Further information

The information you gather will also help you to identify species which are not in this book. If you take careful notes in the field you will be able to use them when referring to other books.

When you are looking for additional information about plants listed in this book, use the Latin names, as the common names often differ from one area to another. The Latin name for each plant is given in the index, along with its most usual common name.

If you are interested in seeing unusual or rare species, you could try natural history museums and botanical gardens in your area.

CHAPTER 1

Wild Flowers

In this Chapter, the flowers are arranged by colour and the description next to each illustration points out the flower's most important identifying features, including its habitat. The pictures in circles next to the main illustrations show close-ups of flowers or sometimes the fruits or seeds of the plant. The plants are not drawn to scale but their average height (measured from the ground) is given. The last line of the description indicates the months during which the plant is in flower.

Identifying wild flowers
Apart from a plant's size and habitat, there are several other important details you should note. Look at the shape and the arrangement of the leaves, the number of petals and stamens, any fruit or seeds that are present and whether the whole plant or any part of it is hairy or smooth. A magnifying glass is useful for examining the plant closely.

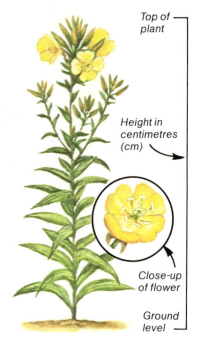

Top of plant

Height in centimetres (cm)

Close-up of flower

Ground level

Looking at fruits and seeds
Fruits are usually best seen after the petals have withered and fallen. Most seeds are contained inside fleshy or dry fruits.

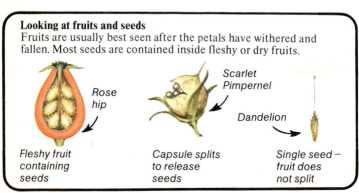

Rose hip

Scarlet Pimpernel

Dandelion

Fleshy fruit containing seeds

Capsule splits to release seeds

Single seed – fruit does not split

Looking at the flower

The stigma and ovary make up the female reproductive organs and the stamens the male ones. Pollen grains from the stamens are received by the stigma and result in the growth of a seed inside the ovary.

In most flowers the reproductive organs are surrounded by sepals and petals. Sepals, which are usually green, make up the outer part of the flower called the calyx. The petals make up the corolla.

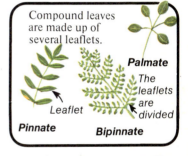

The parts of the flower can be arranged like this, or in other ways.

Looking at leaves

Most leaves consist of a stalk (or petiole) and a flattened blade (or lamina). Leaves come in many shapes and sizes and can be a very important clue when identifying a plant. The main division is into simple leaves, which are undivided, and compound leaves, which are divided into leaflets.

Compound leaves are made up of several leaflets.

Palmate

The leaflets are divided

Leaflet

Pinnate

Bipinnate

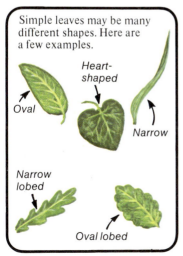

Simple leaves may be many different shapes. Here are a few examples.

Oval

Heart-shaped

Narrow

Narrow lobed

Oval lobed

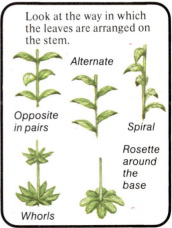

Look at the way in which the leaves are arranged on the stem.

Alternate

Opposite in pairs

Spiral

Whorls

Rosette around the base

7

Look for these flowers in damp places, such as ditches, marshes and water meadows.

Lesser Celandine ▶

A small, creeping plant with glossy, heart-shaped leaves. Shiny yellow flowers. Look in damp shady woods and waysides. 7 cm tall. March-May.

◀ Alternate-leaved Golden Saxifrage

Small plant with round, toothed leaves, and greenish yellow flowers. Look in wet places. 7 cm tall. April-July.

Each flower has four yellow sepals

Creeping Buttercup ▶

Look for the long runners near the ground. Hairy, deeply-divided leaves. Shiny yellow flowers. Common weed of grassy places. May-Aug.

Runner

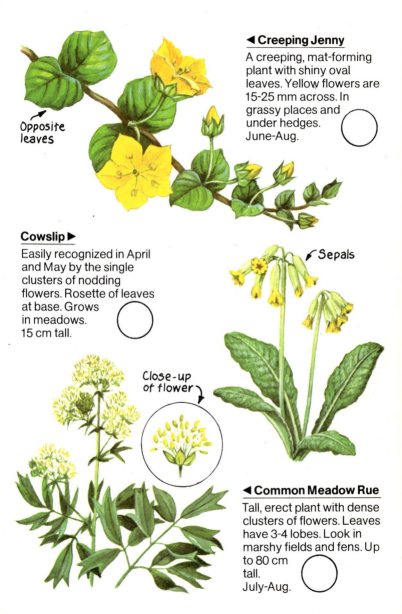

◄ Creeping Jenny

A creeping, mat-forming plant with shiny oval leaves. Yellow flowers are 15-25 mm across. In grassy places and under hedges. June-Aug.

Opposite leaves

Cowslip ►

Easily recognized in April and May by the single clusters of nodding flowers. Rosette of leaves at base. Grows in meadows. 15 cm tall.

Sepals

Close-up of flower

◄ Common Meadow Rue

Tall, erect plant with dense clusters of flowers. Leaves have 3-4 lobes. Look in marshy fields and fens. Up to 80 cm tall. July-Aug.

Look for these flowers, and those on page 11, in woods, hedgerows and heaths.

Herb Bennet or Wood Avens ▶

Fruits have hooks which catch on clothes and animals' fur. Woods, hedges and shady places. Up to 50 cm tall. June-Aug.

Cluster of fruits

◀ Yellow Pimpernel

Like Creeping Jenny, but smaller, with more pointed leaves. Slender trailing stems. The flowers close in dull weather. Woods and hedges. May-Sept.

Barberries can be used to make jam

Barberry ▶

A shrub with spiny branches. Bees visit the drooping flowers. Look for the red berries. Hedges and scrubland Up to 1 m tall. May-June.

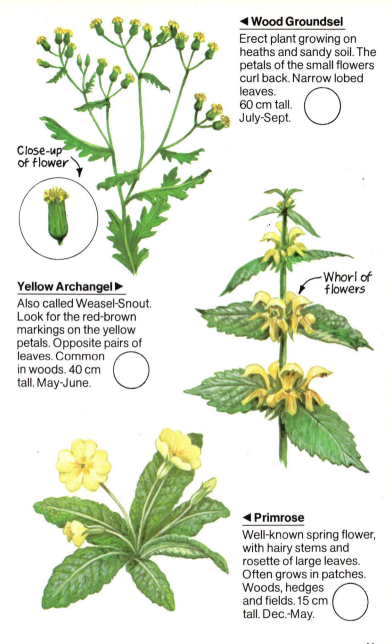

◄ Wood Groundsel

Erect plant growing on heaths and sandy soil. The petals of the small flowers curl back. Narrow lobed leaves.
60 cm tall.
July-Sept.

Close-up of flower

Yellow Archangel ►

Also called Weasel-Snout. Look for the red-brown markings on the yellow petals. Opposite pairs of leaves. Common in woods. 40 cm tall. May-June.

Whorl of flowers

◄ Primrose

Well-known spring flower, with hairy stems and rosette of large leaves. Often grows in patches. Woods, hedges and fields. 15 cm tall. Dec.-May.

Look for these flowers, and those on page 13, in open grassy places, such as heaths and commons.

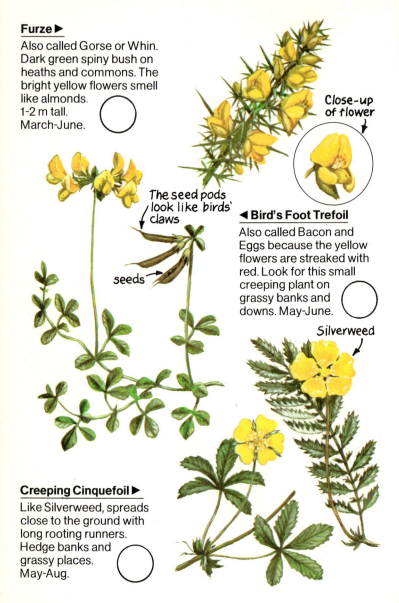

Furze ▶

Also called Gorse or Whin. Dark green spiny bush on heaths and commons. The bright yellow flowers smell like almonds.
1-2 m tall.
March-June.

Close-up of flower

The seed pods look like birds' claws

seeds

◀ Bird's Foot Trefoil

Also called Bacon and Eggs because the yellow flowers are streaked with red. Look for this small creeping plant on grassy banks and downs. May-June.

Silverweed

Creeping Cinquefoil ▶

Like Silverweed, spreads close to the ground with long rooting runners. Hedge banks and grassy places.
May-Aug.

◄ Common St John's Wort
Look for see-through dots
on the narrow oval leaves
and black dots on the
petals and sepals. Damp
grassy places.
60 cm tall.
June-Sept.

Woad ►
Look for the hanging pods
on this tall, erect plant. The
leaves were once boiled
to make a blue dye.
Waysides and dry
places. 70 cm
tall. June-Sept.

Seed
pod

Dandelion
"clock"

Close-up
of fruit

◄ Dandelion
Common weed with rosette
of toothed leaves. The
flowers close at night.
Look for the "clock" of
downy white fruits.
Waysides. 15 cm
tall. March-June.

Stonecrop ▶

Also called Wallpepper.
Mat-forming plant with
star-shaped flowers. The
thick fleshy leaves have
a peppery taste.
Dunes, shingle and
walls. June-July.

Close-up
of a flower

Leaves

◀ Purslane

A low spreading plant with
red stems. The fleshy oval-
shaped leaves are in
opposite pairs. A weed of
fields and waste
places.
May-Oct.

Close-up
of a flower

Golden Rod ▶

Erect plant with flowers on
thin spikes. Leaves are
narrower and more pointed
near top of plant. Woods,
banks and cliffs.
40 cm tall.
July-Sept.

Leaves
broader
near bottom
of plant

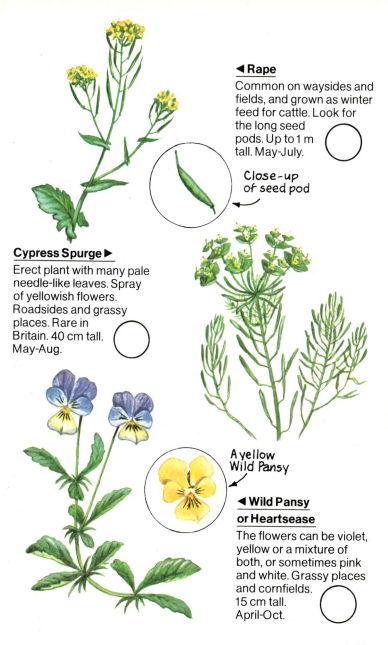

◀ Rape

Common on waysides and fields, and grown as winter feed for cattle. Look for the long seed pods. Up to 1 m tall. May-July.

Close-up of seed pod

Cypress Spurge ▶

Erect plant with many pale needle-like leaves. Spray of yellowish flowers. Roadsides and grassy places. Rare in Britain. 40 cm tall. May-Aug.

A yellow Wild Pansy

◀ Wild Pansy or Heartsease

The flowers can be violet, yellow or a mixture of both, or sometimes pink and white. Grassy places and cornfields. 15 cm tall. April-Oct.

Cornflower ▶

Also called Bluebottle.
Erect plant with greyish
downy leaves and a blue
flower head. Cornfields
and waste places.
40 cm tall.
July-Aug. Rare.

Seed
pod

◀ Larkspur

Slender plant with divided
feathery leaves. The
flowers have a long spur.
Cultivated land.
50 cm tall.
June-July.

Spur

Flower
bud

Lesser Periwinkle ▶

Creeps along the ground
with long runners, making
leafy carpets. Shiny oval
leaves. Woods and hedges.
Flower stems up
to 15 cm tall.
Feb.-May.

Runner

Runner

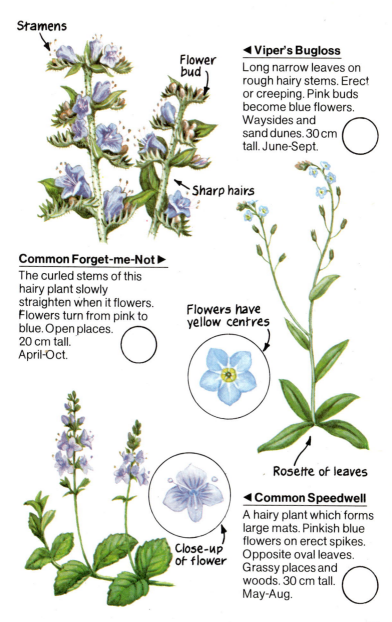

Stamens

Flower bud

Sharp hairs

◀ Viper's Bugloss
Long narrow leaves on rough hairy stems. Erect or creeping. Pink buds become blue flowers. Waysides and sand dunes. 30 cm tall. June-Sept.

Common Forget-me-Not ▶
The curled stems of this hairy plant slowly straighten when it flowers. Flowers turn from pink to blue. Open places. 20 cm tall. April-Oct.

Flowers have yellow centres

Rosette of leaves

Close-up of flower

◀ Common Speedwell
A hairy plant which forms large mats. Pinkish blue flowers on erect spikes. Opposite oval leaves. Grassy places and woods. 30 cm tall. May-Aug.

Look for the flowers shown on this page in damp places.

Common Monkshood ▶

Also called Wolfsbane.
Notice hood on flowers
and the deeply-divided
leaves. Near streams and
in damp woods.
70 cm tall.
June-Sept.

*Flower is
shaped like a
monk's hood*

◀ Brooklime

Creeping plant with erect
reddish stems. Shiny oval
leaves in opposite pairs.
Used to be eaten in
salads. Wet
places. 30 cm tall.
May-Sept.

Bugle ▶

Creeping plant with erect
flower spikes. Purplish
stem is hairy on two
sides. Forms carpets in
damp woods.
10-20 cm tall.
May-June.

*Close-up of
bugle-shaped
flower*

Flower

Fruiting head

◄ Sea Holly
A stiff, spiny plant with grey-blue leaves and round flower heads. Look for it on sandy and shingle beaches.
50 cm tall.
July-Aug.

Meadow Clary or Meadow Sage ►
Hairy stem with wrinkled leaves mostly at the base of the plant.
Grassy places.
40 cm tall.
June–July.

◄ Bluebell
Also called Wild Hyacinth. Narrow, shiny leaves and clusters of nodding blue flowers. Forms thick carpets in woods.
30 cm tall.
April-May.

Close-up of fruit

19

Look for the flowers shown on this page in woods or hedges.

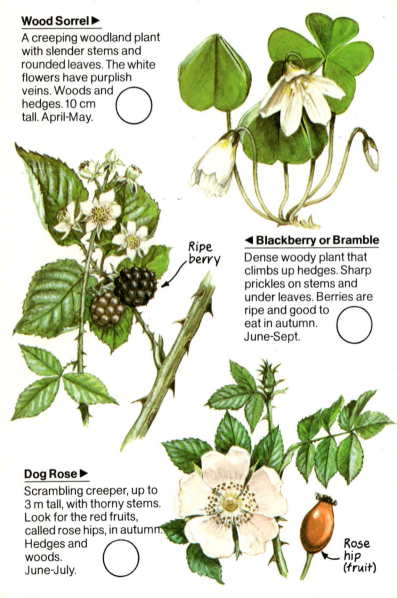

Wood Sorrel ▶

A creeping woodland plant with slender stems and rounded leaves. The white flowers have purplish veins. Woods and hedges. 10 cm tall. April-May.

Ripe berry

◀ Blackberry or Bramble

Dense woody plant that climbs up hedges. Sharp prickles on stems and under leaves. Berries are ripe and good to eat in autumn. June-Sept.

Dog Rose ▶

Scrambling creeper, up to 3 m tall, with thorny stems. Look for the red fruits, called rose hips, in autumn. Hedges and woods. June-July.

Rose hip (fruit)

◀ Bistort

Also called Snakeweed.
Forms patches. Leaves are
narrow. Flowers in spikes.
In meadows, often near
water.
40 cm tall.
June-Oct.

Greater Bindweed ▶

Look for the large pink or
white funnel-shaped
flowers. Climbs walls and
hedges in waste places.
Leaves are shaped like
arrowheads.
3 m high.
July-Sept.

Flower
bud

◀ Red Helleborine

Upright plant with pointed
leaves and a fleshy stem.
Rare plant, protected by
law. Woods and shady
places. Up to
40 cm tall.
May-June.

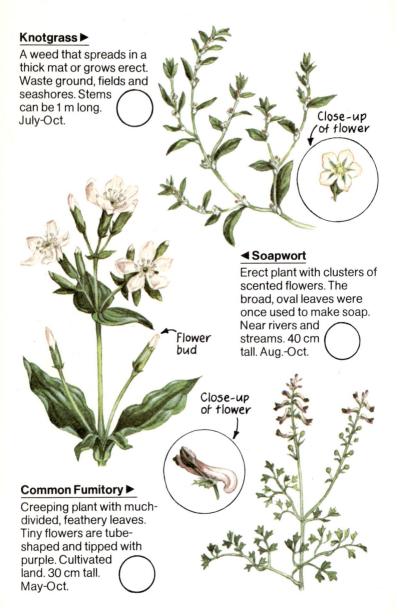

Knotgrass ▶

A weed that spreads in a thick mat or grows erect. Waste ground, fields and seashores. Stems can be 1 m long. July-Oct.

Close-up of flower

◀ Soapwort

Erect plant with clusters of scented flowers. The broad, oval leaves were once used to make soap. Near rivers and streams. 40 cm tall. Aug.-Oct.

Flower bud

Close-up of flower

Common Fumitory ▶

Creeping plant with much-divided, feathery leaves. Tiny flowers are tube-shaped and tipped with purple. Cultivated land. 30 cm tall. May-Oct.

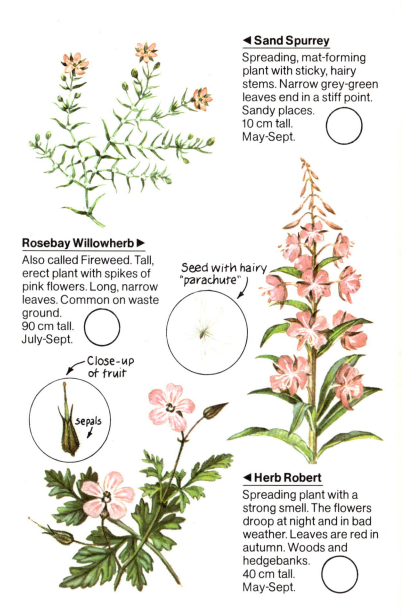

◄ Sand Spurrey

Spreading, mat-forming
plant with sticky, hairy
stems. Narrow grey-green
leaves end in a stiff point.
Sandy places.
10 cm tall.
May-Sept.

Rosebay Willowherb ►

Also called Fireweed. Tall,
erect plant with spikes of
pink flowers. Long, narrow
leaves. Common on waste
ground.
90 cm tall.
July-Sept.

Seed with hairy
"parachute"

Close-up
of fruit

sepals

◄ Herb Robert

Spreading plant with a
strong smell. The flowers
droop at night and in bad
weather. Leaves are red in
autumn. Woods and
hedgebanks.
40 cm tall.
May-Sept.

Look for these flowers on heaths and moors.

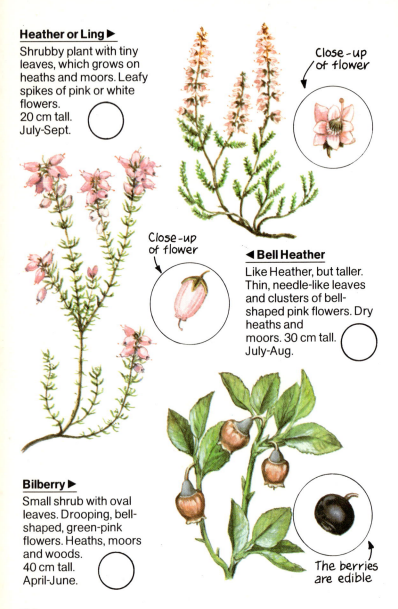

Heather or Ling ▶

Shrubby plant with tiny leaves, which grows on heaths and moors. Leafy spikes of pink or white flowers.
20 cm tall.
July-Sept.

Close-up of flower

Close-up of flower

◀ Bell Heather

Like Heather, but taller. Thin, needle-like leaves and clusters of bell-shaped pink flowers. Dry heaths and moors. 30 cm tall.
July-Aug.

Bilberry ▶

Small shrub with oval leaves. Drooping, bell-shaped, green-pink flowers. Heaths, moors and woods.
40 cm tall.
April-June.

The berries are edible

Look for these flowers in dry, grassy places.

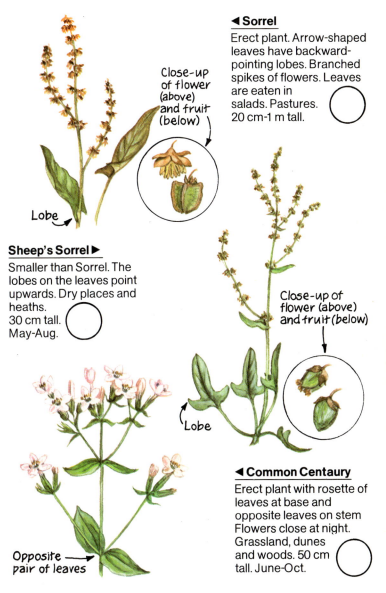

◀ Sorrel

Erect plant. Arrow-shaped leaves have backward-pointing lobes. Branched spikes of flowers. Leaves are eaten in salads. Pastures. 20 cm-1 m tall.

Close-up of flower (above) and fruit (below)

Lobe

Sheep's Sorrel ▶

Smaller than Sorrel. The lobes on the leaves point upwards. Dry places and heaths. 30 cm tall. May-Aug.

Close-up of flower (above) and fruit (below)

Lobe

◀ Common Centaury

Erect plant with rosette of leaves at base and opposite leaves on stem Flowers close at night. Grassland, dunes and woods. 50 cm tall. June-Oct.

Opposite pair of leaves

Ragged Robin ▶

Flowers have ragged pink petals. Erect plant with a forked stem and narrow, pointed leaves. Damp meadows, marshes and woods.
30-70 cm tall.
May-June.

A bract is a kind of small leaf near the flower

Grooved stem

◀ Knapweed or Hard-head

Erect plant with brush-like pink flowers growing from black bracts. Grassland and waysides.
40 cm tall.
June-Sept.

Hemp Agrimony ▶

Tough, erect plant with downy stem. Grows in patches in damp places. Attracts butterflies.
Up to 120 cm tall.
July–Sept.

Whorl of leaves

◄ Deptford Pink

The clusters of bright pink flowers close in the afternoon. Pointed, opposite leaves. Very rare in Britain. Sandy places. 40 cm tall. July-Aug.

Close-up of flower

Blood-red Geranium or Bloody Cranesbill ►

Bushy plant with erect or trailing stems. Deeply divided leaves are round and hairy. Hedgerows. 30 cm tall. June-Aug.

Fruit

Seed pod

◄ Red Campion

Erect plant with a hairy, sticky stem and pointed, oval leaves in opposite pairs. Woodland. 60 cm tall. May-June.

◄ Early Purple Orchid

Erect plant with dark spots on the leaves. Smells like cats. Look for the hood and spur on the flower. Woods and copses. Up to 60 cm tall. June-Aug.

Foxglove ►

Erect plant with tall spike of tube-shaped flowers, drooping on one side of the stem. Large oval leaves. Open woods. Up to 1.5 m tall. June-Sept.

Policeman's Helmet (closely related to Touch-me-not Balsam)

Touch-me-not Balsam

◄ Policeman's Helmet

Also called Jumping Jack. Flowers look like open mouths. Ripe seed pods explode, scattering seeds when touched. Streams. Up to 2 m tall. July-Oct.

Look for the flowers shown on this page in woods or hedgerows.

Bats-in-the-Belfry ▶
Erect hairy plant with large toothed leaves. Flowers on leafy spikes point upwards. Hedges, woods and shady places. 60 cm tall. July-Sept.

Tendril

◀ Tufted Vetch
Scrambling plant with clinging tendrils. Climbs up hedgerows. Look for the brown seed pods in late summer. Flowers 10 mm across. June-Sept.

Spur

Sepals

Bud

Common Dog Violet ▶
Creeping plant with rosettes of heart-shaped leaves. Look for the pointed sepals and short spur on the flower. Woods. 10 cm tall. April-June.

Look in fields and other grassy places for these flowers.

Pasque Flower ▶

Very rare in the wild, but grows in gardens. Hairy feathery leaves. Purple or white flowers have yellow anthers. Dry grassy places. 10 cm tall. April-May.

Field Scabious is a similar species

Devil's Bit Scabious

◀ Devil's Bit Scabious

Erect plant with narrow, pointed leaves. Flowers are pale to dark purple. Round flower heads. Wet grassy places. 15-30 cm tall. June-Oct.

Note its lobed leaves

Note its entire leaves

Fritillary or Snake's Head ▶

Drooping flowers are checkered with light and dark purple. Varies from white to dark purple. Damp meadows. 10 cm tall. May.

You may see these flowers on old walls.

◄ Ivy-leaved Toadflax

Weak, slender stalks trail on old walls. Look for the yellow lips on the mauve flowers. Flowers 10 mm across. Shiny, ivy-shaped leaves. May-Sept.

Spur

Houseleek ►

A rosette plant with thick fleshy leaves. Dull red spiky petals. Does not flower every year. Old walls and roofs. 30-60 cm tall. June-July.

Rosette of leaves

The stalk, with flowers, does not appear very often. Usually you will see only the rosette.

Fruits

◄ Snapdragon

Erect plant with spike of flowers. Long, narrow leaves. Pouch-like flowers are yellow inside. Old walls, rocks and gardens. 40 cm tall. June-Sept.

Look for these flowers in cornfields and on farmland.

Scarlet Pimpernel ▶

Grows along the ground.
Flowers close in bad
weather. Black dots under
the pointed oval leaves.
Cultivated land.
15 cm tall.
June-Aug.

Flowers may
also be blue

◀ Poppy

Erect plant with stiff hairs
on stem. Soft red flowers
have dark centres. Round
seed pod. Cornfields and
waste ground. Up
to 60 cm tall.
June-Aug.

Seed pod

Flower
bud

Seed
pod

Long-headed Poppy ▶

Like Poppy, but flowers
are paler and do not have
dark centres. Pod is long
and narrow. Cornfields
and waste ground.
Up to 45 cm tall.
June-Aug.

Summer Pheasant's Eye (not in Britain) is a similar species

◄ Pheasant's Eye

Rare cornfield weed, with finely divided feathery leaves. The red flowers have black centres. 20 cm tall. May-Sept.

Sweet William ►

Tough, narrow leaves and flat flower cluster. Mountain pastures and cultivated land in Europe. Gardens only in Britain. 60 cm tall. May-June.

Close-up of flower

◄ Wood Woundwort

The leaves were once used to dress wounds. Spikes of dark red and white flowers in whorls. Smells strongly. Woods. 40 cm tall. June–Aug.

33

The flowers on these two pages can be found in woodlands, quite early in the year.

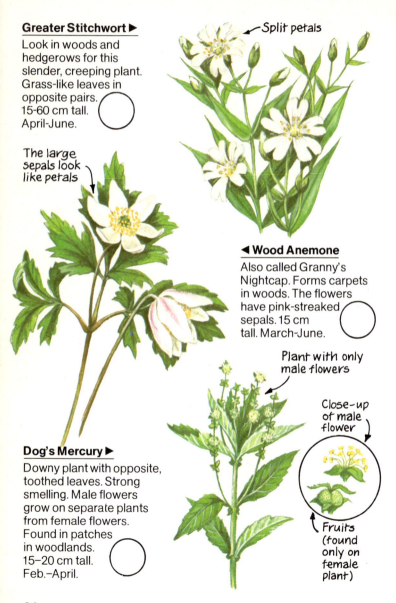

Greater Stitchwort ▶

Look in woods and hedgerows for this slender, creeping plant. Grass-like leaves in opposite pairs. 15-60 cm tall. April-June.

Split petals

The large sepals look like petals

◀ Wood Anemone

Also called Granny's Nightcap. Forms carpets in woods. The flowers have pink-streaked sepals. 15 cm tall. March-June.

Plant with only male flowers

Close-up of male flower

Dog's Mercury ▶

Downy plant with opposite, toothed leaves. Strong smelling. Male flowers grow on separate plants from female flowers. Found in patches in woodlands. 15–20 cm tall. Feb.–April.

Fruits (found only on female plant)

◄ Ramsons or Wood Garlic

Smells of garlic. Broad, bright green leaves grow from a bulb. Forms carpets in damp woods, often with Bluebells.
10-25 cm tall.
April-June.

Notice the long veins that run from one end of the leaf to the other

Lily-of-the-Valley ►

Grows in dry woods. Broad, dark green leaves and sweet-smelling flowers. Red berries in summer. Also a garden plant. 20 cm tall. May-June.

Berry

◄ Snowdrop

Welcomed as the first flower of the new year. Dark green, narrow leaves. Nodding white flowers. Woods.
20 cm tall.
Jan.-March.

Look for these flowers in hedges or woods.

Jack-by-the-Hedge or Garlic Mustard ▶

Erect plant with heart-shaped, toothed leaves. Smells of garlic. Common in hedges. Up to 1.2 m tall. April-June.

Seed pods

Fruits are smaller than garden strawberries

◀ Wild Strawberry

Small plant with long, arching runners and oval, toothed leaves in threes. Sweet red fruits, covered with seeds. Woods and scrubland. April-July.

Tendril

Wild Pea ▶

Very rare, scrambling plant with grey-green leaves. The seeds, or peas, are inside the pods. Climbs on thickets and hedges. Up to 2.5 m high. June–Aug.

Pod

Look for these flowers in hedges and waysides.

Close-up of a female flower

◀ White Bryony

Climbs up hedges with spiral tendrils. The red berries appear in August and are poisonous. Large underground stems, called tubers. Up to 4 m tall. June.

Tendril

Berries

Cow Parsley ▶

Also called Lady's Lace. Look for the ribbed stem, feathery leaves and white flower clusters. Hedge banks and ditches. Up to 1 m tall. May-June.

Close-up of a flower

Fruit

Close-up of a flower

◀ Hedge Parsley

Like Cow Parsley, but with a stiff, hairy stem. Look for the prickly purple fruits. Cornfields and roadsides. 60 cm tall. April-May.

Fruit

These flowers can be found in or near fresh water (streams, ponds, etc.).

Meadowsweet ▶

Clusters of sweet smelling flowers. Grows in marshes, water meadows, and also near ditches at the side of the road. Up to 80 cm tall. May–Sept.

Undersides of leaves are silvery-grey

◀ Triangular-stalked Garlic or Three-cornered Leek

Smells of garlic. Drooping flowers. In damp hedges and waste places. Not in Britain. 40 cm tall. June-July.

The flower stem is three-sided

Water surface

Underwater leaves are longer and thinner

Floating Water Plantain ▶

Water plant with oval leaves and white flowers on the water surface. Look for it in canals and still water. Flowers 12-15 cm across. May-Aug.

These flowers can be found in or near fresh water (streams, ponds, etc.).

◄ Water Crowfoot
Water plant whose roots are anchored in the mud at the bottom of ponds and streams. Flowers (10–20 mm across) cover the water surface. May–June.

These leaves are on the water surface

Fine, underwater leaves

Water Soldier ►
Under water except when it flowers. Long saw-like leaves then show above the surface. Flowers 30-40 mm across. Ponds, canals, ditches. June-Aug.

Bud

◄ Frogbit
Rises to the surface in spring, and spreads with long runners. Shiny round leaves grow in tufts. Flowers 20 mm across. Canals and ponds. July-Aug.

Runner

Look for these flowers in fields and other grassy places.

Wild Carrot ▶

Dense clusters of white flowers with a purple flower in the centre. Erect, hairy stem with feathery leaves. Grassy places, often near coast.
60 cm tall.
July–Aug.

Clusters of small flowers

Close-up of a single flower

Bracts

Fruit

Cluster of fruits

◀ Hogweed or Keck

Very stout, hairy plant with huge leaves on long stalks. Flowers are in clusters. Grassy places and open woods.
Up to 1 m tall.
June-Sept.

Close-up of single flower

Fruit

Single flower

Corky-fruited Water Dropwort ▶

Erect plant with large, much-divided, feathery leaves. Clusters of flowers. Meadows.
60 cm tall.
June-Aug.

Fruit

Look for these flowers in fields and other grassy places.

White petals are sometimes tinged with pink

◀ Daisy

Small plant with rosette of leaves at base. Flowers close at night and in bad weather. Very common on garden lawns.
10 cm tall.
Jan.-Oct.

White or Dutch Clover ▶

Creeping plant often grown for animal feed. Look for the white band on the three-lobed leaves. Attracts bees.
10-25 cm tall.
April-Aug.

White band

Runner

Look for the divided petals

◀ Field Mouse-ear Chickweed

Creeping plant with erect stems. Narrow, downy leaves. Grassy places.
10 cm tall.
April-Aug.

Look for these flowers on cultivated land, waste land and waysides.

Nettle ▶
The toothed leaves are covered with stinging hairs. Dangling green-brown flowers. Used to make beer and tea. Common. Up to 1 m tall. June-Aug.

Cluster of flowers

Single flower

Fruit

Close-up of flower

◀ Pigweed or Common Amaranth
Erect hairy plant with large oval leaves. Large spikes of green tufty flowers. Look for it on cultivated land. 50 cm tall. July-Sept.

Close-up of flower

Common Orache ▶
An erect weed with a stiff stem and toothed leaves, both dusty grey. Cultivated land or waste places. Up to 90 cm tall. Aug.-Sept.

Look for these flowers on cultivated land, waste land and waysides.

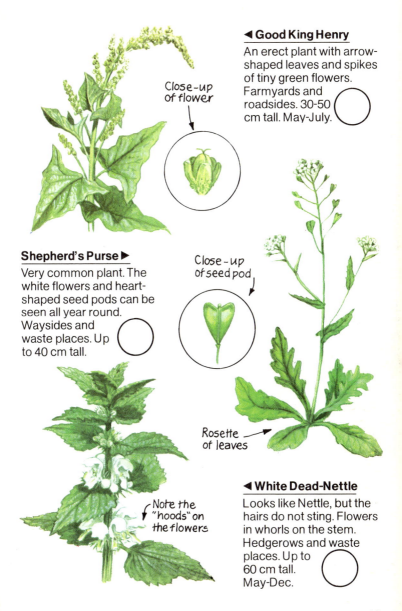

◀ Good King Henry

An erect plant with arrow-shaped leaves and spikes of tiny green flowers. Farmyards and roadsides. 30-50 cm tall. May-July.

Close-up of flower

Shepherd's Purse ▶

Very common plant. The white flowers and heart-shaped seed pods can be seen all year round. Waysides and waste places. Up to 40 cm tall.

Close-up of seed pod

Rosette of leaves

Note the "hoods" on the flowers

◀ White Dead-Nettle

Looks like Nettle, but the hairs do not sting. Flowers in whorls on the stem. Hedgerows and waste places. Up to 60 cm tall. May-Dec.

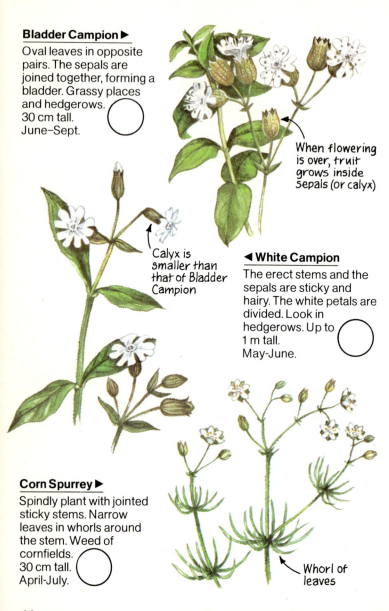

Bladder Campion ▶

Oval leaves in opposite pairs. The sepals are joined together, forming a bladder. Grassy places and hedgerows. 30 cm tall. June–Sept.

When flowering is over, fruit grows inside sepals (or calyx)

Calyx is smaller than that of Bladder Campion

◀ White Campion

The erect stems and the sepals are sticky and hairy. The white petals are divided. Look in hedgerows. Up to 1 m tall. May-June.

Corn Spurrey ▶

Spindly plant with jointed sticky stems. Narrow leaves in whorls around the stem. Weed of cornfields. 30 cm tall. April-July.

Whorl of leaves

◄ Chickweed

Mat-forming plant with stems that can grow up to 40 cm tall. You can see the small flowers all year round. Common weed in fields, gardens.

Black Nightshade ►

Shrubby weed of cultivated ground. Shiny oval leaves. Petals fold back to show yellow anthers. The berries are poisonous. 20 cm tall. July-Sept.

Anthers

Berries

Whorl of leaves

Fruit

◄ Goosegrass or Common Cleavers

Scrambling plant. The prickly stems stick to clothes and animal fur. Hedges. 60 cm tall. June-Sept.

Look for these flowers in grassy places, on waste or cultivated ground.

Ribwort Plantain or Cocks and Hens ▶

Tough plant with narrow, ribbed leaves. Green-brown spikes of flowers have white anthers. Common. 20 cm tall. April-Aug.

Anthers

Anthers are mauve at first, changing to yellow

Anthers

◀ Greater Plantain or Ratstail

Broad-ribbed leaves in a rosette close to the ground. All kinds of cultivated land. 15 cm tall. May-Sept.

Anthers

Hoary Plantain ▶

Rosette plant with oval, ribbed leaves. Fine hairs on stem. White flowers have purple anthers. Common in grassy places. 7-15 cm tall. May-Aug.

Look for these flowers on grassy or waste ground.

◀ Yarrow

Common plant with rough stem and feathery leaves. Flat-topped clusters of flowers. Smells sweet. Was once used to heal wounds. 40 cm tall. June-Aug.

Wild Chamomile or Scented Mayweed ▶

Erect plant with finely divided leaves. The petals fold back. Waste places everywhere. 15-40 cm tall. June-July.

◀ Ox-eye Daisy or Marguerite

Erect plant with rosette of toothed leaves and large daisy-like flowers. Roadsides and grassy places. Up to 60 cm tall. June-Aug.

CHAPTER 2

Trees

The illustrations in this Chapter will help you identify a tree at any time of year. For each tree, the leaf, the bark, and the shape of the mature tree in full leaf are shown. If the tree is deciduous, the shape of the tree in winter is also shown. Flowers and fruits, including cones, are illustrated if they will help you identify the tree. The average height of a mature tree is written next to each illustration.

Broadleaved trees
Most broadleaved trees are deciduous. They have broad, flat leaves that change from green to shades of red and gold in autumn and are then shed. In spring, new leaves unfurl from the winter buds.

The seeds of broadleaved trees are enclosed in fruits. The fruits vary from dry and nut-like to soft and fleshy, and they take a variety of forms.

The wood of broadleaved trees is called hardwood.

Conifers
Most coniferous trees are evergreens – their leaves do not change colour in autumn and they are not shed all at once. Some conifers, such as pine and cedar, have long narrow leaves called needles. Others, such as cypress, have flattened leaves which lie close to the twig.

The seeds of conifers are produced in fruits called cones.

The wood of coniferous trees is called softwood.

Conifers include the largest and oldest of all known living organisms.

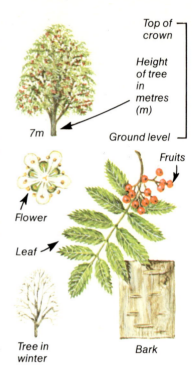

Top of crown

Height of tree in metres (m)

7m — Ground level

Fruits

Flower

Leaf

Tree in winter

Bark

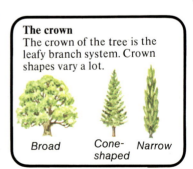

The crown
The crown of the tree is the leafy branch system. Crown shapes vary a lot.

Broad

Cone-shaped

Narrow

Broadleaved trees

Leaves
Here are some examples of leaf shapes:

Oval

Deeply lobed

Compound Palmate

Narrow

Flowers and fruits

Blossom

Apple

Crab Apple blossom develops into fleshy fruits containing seeds.

Clusters of small flowers

Pod

Laburnum flowers develop into pods containing seeds.

Conifers

Leaves
Here are some examples of leaf types:

Rosette-like clusters of needles

Short single needles

Small bunches of long needles

Scale-like leaves covering twigs

Cones

Conifers have tiny, flower-like buds that develop into cones containing seeds.

Larch "flowers"

Larch cone

Small bud

Short, paired needles

Long, bare trunk is red near top of tree

35m

▲ Scots Pine

Short, blue-green, paired needles. Small pointed buds. Upper bark red, but grey and furrowed below. Young tree pointed, becoming flat-topped with age.

Green, pointed cone turns brown in second year

Bark flakes off in "plates"

Cones are on tree for several years

Long, paired needles

Long bud

22m

▲ Maritime Pine

Long, stout, grey-green, paired needles. Long, spindle-shaped buds. Long, shiny brown cones in clusters. Rugged bark on long, bare trunk.

Young shoots, seen in early summer

Cones later turn brown with age

20m

▲ Stone Pine

Long, dark green, paired needles. Buds small. Large broad cones with edible seeds. Umbrella-shaped tree with a flat top. Mediterranean.

Paired needles

Bud

Prickly scales

Young shoot

23m

▲ Shore Pine

Yellow-green, paired needles on twisted shoots. Sticky, bullet-shaped buds. Small cones in clusters. Scaly bark. Tall, narrow, fast-growing tree.

Paired needles

Bud

Young shoot, seen in early summer

Paired needles

36m

▲ Corsican Pine

Long, dark green, paired needles. Onion-shaped buds. Large, lop-sided, brown cones. Blackish bark. Tall tree with regular branchings.

Bud

Cones take two years to ripen

(Rare in Britain)

10m

▲ Aleppo Pine

Bright green, paired needles. Small round buds. Cones usually in groups of two or three. Small, round-topped tree. Common in Mediterranean.

Paired needles

Young shoot

Shiny, reddish cones stay on tree for many years

Bud

Young shoot, seen in early summer

Lower branches usually touch the ground

17m

Needles in fives

Bark is rugged and scaly

▲ Swiss Stone Pine

Dense, stiff needles in fives. Small, pointed, sticky buds. Egg-shaped cones, with edible seeds, ripen and fall in third year. Cone-shaped tree.

Bud

Heavily-branched, broad crown

Needles in threes

Young shoot

30m

Cones uneven at base

Bud

▲ Monterey Pine

Slender, grass green needles in threes. Large, pointed, sticky buds. Cones squat, growing flat against branches, staying on tree for many years.

53

Buds

Cone scales
are tightly
closed

30m

▲ Norway Spruce

Prickly, dark green needles.
Small brown buds. Peg-like
bumps left on brown twigs
when needles are pulled off.
Cone-shaped
tree. Used as
Christmas tree.

Cones have papery
scales with crinkled
edges

35m

Buds

▲ Sitka Spruce

Very prickly, blue-green
needles. Plump yellow
buds. Small knobs left on
yellow twigs when needles
fall off. Narrow,
cone-shaped
trees.

Grey, scaly bark
flakes off in
"plates"

Fine branches

Straw-coloured twigs

38m

▲ European Larch

Bunches of soft, light green needles, which fall in winter, leaving small knobs on twigs. Female flowers are reddish. Small egg-shaped cones.

Tree is deciduous

Stout branches

Edges of scales turn backwards

35m

▲ Japanese Larch

Bunches of blue-green needles, falling in winter. Orange twigs. Female flowers are pinkish-green. Small, rosette-like cones.

Tree is deciduous

Young cones
are green,
older ones
plum-coloured

▲ Nootka Cypress

Fern-like sprays of dull
green, scale-like leaves
grow on either side of
twig. Each scale of plum-
coloured cones
has a prickle.
Cone-shaped
crown.

Flat-topped
crown

Large, upright
cones

Needles have
notched tips

Tall,
narrow
tree

40m

Bracts
showing

▲ European Silver Fir

Flat single needles are
green above and silvery
below. Flat, round scars are
left on twigs when needles
drop off. Cones
shed their scales
when ripe.

25m

30m

▲ Greek Fir

Shiny green, spiny-tipped needles all round twig. Tall, narrow cones shed scales to leave bare spike on tree. Common in parks.

Pointed tip

Bark flakes off in "plates"

28m

▲ Spanish Fir

Short, blunt, blue-grey needles all round twig. Cylindrical, upright cones fall apart on tree. Found only in gardens in Britain.

Blunt tip

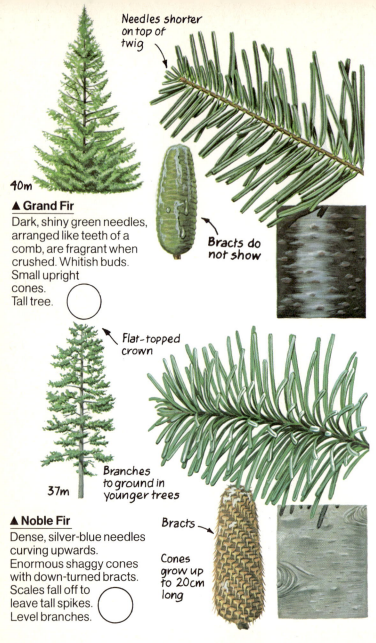

Needles shorter on top of twig

40m

▲ Grand Fir

Dark, shiny green needles, arranged like teeth of a comb, are fragrant when crushed. Whitish buds. Small upright cones. Tall tree.

Bracts do not show

Flat-topped crown

Branches to ground in younger trees

37m

▲ Noble Fir

Dense, silver-blue needles curving upwards. Enormous shaggy cones with down-turned bracts. Scales fall off to leave tall spikes. Level branches.

Bracts

Cones grow up to 20cm long

Beech-like bud

Needles are parted on twig

Bracts

40m

▲ Douglas Fir
Soft fragrant needles.
Long-pointed, copper-
brown buds. Light brown,
hanging cones with three-
pointed bracts.
Old bark is thick
and corky.

Leader droops

Young cones are green, older ones are brown

Cone has a few rounded scales

Tips of branches droop

35m

▲ Western Hemlock
Needles various lengths,
green above and silver
below. Small cones on
shoot tips. Smooth, brown
scaly bark.
Branch tips and
top shoot droops.

Flattened needles

Leading shoot is upright

Cone

30m

▲ Western Red Cedar

Flattened sprays of scale-like leaves covering twigs. Small, flower-shaped cones. Smooth, finely-furrowed bark.

Leaves are dark shiny green above and whitish below

Cone scales are attached at the centre

25m

▲ Lawson Cypress

Sprays of fine, scale-like leaves, green and other colours. Small round cones. Smooth, reddish bark. Leader shoot is often forked.

Spray of scale-like leaves

15m

Cones are shiny
pale green at
first, dull grey
when older

Leaves are
smaller than
those of
Monterey
Cypress

▲ Italian Cypress

Small, dark, dull-green,
scale-like leaves, closely
pressed to stem. Large,
grey, rounded cones. An
upright, narrow-
crowned tree.
Mainly ornamental.

Leaves are
lemon-scented
when crushed

25m

Knob

▲ Monterey Cypress

Dense sprays of small,
scale-like leaves. Large,
purplish-brown, rounded
cones with knob on scales.
Column-shaped
when young, flat-
topped when old.

Peeling
bark

Foliage not
dense

20m

▲ Swamp Cypress

Soft, feathery, light green
needles drop in winter
leaving orange twigs.
Round, purplish-
brown cones.
Triangular-shaped
crown.

Leaves appear
very late

Reddish-brown
spiralled bark,
often
peeling

Tree is
deciduous

Lower branches
touch the
ground

30m

▲ Leyland Cypress

Sprays of dense, bright
green, scale-like leaves.
Round, grey-brown cones
are rare. Thick, column-
like shape. Plants grow
fast from
cuttings. Often
a hedge.

Reddish-brown,
furrowed bark

Cone

30m

▲ Japanese Red Cedar

Long, bright green, spiky
needles curve away from
twig. Round, spiky, green
cones ripening to brown.
Red-brown, peeling bark.
Tall, narrow,
cone-shaped
tree.

Berry-like
cone

Sharp needles

6m

▲ Juniper

Sharp, blue-green needles
in threes with white band on
upper surface. Berry-like
cones turning purplish-
black in second
year. Often
a shrub.

Needles smell
strongly when
crushed

Wide-spreading branches

15m

▲ Yew

Broad needles, dark green above and yellowish-green below, parted on twig. Red, berry-like fruits. Orange-brown flaking bark. Short stout trunk. Can be a hedge.

Leaves and berries are poisonous

Needles turn reddish in autumn

Leaves parted on twig

Long-stalked cones are rare

20m

▲ Dawn Redwood

Soft, light green needles, similar to Swamp Cypress but larger, drop in winter. Young bark is orange and flaking, furrowed in older trees.

Tree is deciduous

33m

▲ Coast Redwood

Hard, sharp-pointed
needles, dark green above
and white-banded below.
Small, round cones. Thick,
reddish, spongy
bark. Tall
tree.

**Needles parted
on either side
of twig** →

38m

▲ Wellingtonia

Deep green, scale-like,
pointed leaves. Long-
stalked, round, corky
cones. Soft, thick, deeply-
furrowed bark.
Tall tree with
upswept branches.

**Foliage hanging
from upswept
branches** ←

**Diamond-shaped
cone scales
wrinkle when
they ripen** ↙

Leaves are blue-green in the common garden variety, dark green in the wild

Sunken top

▲ Atlas Cedar

Dark green needles in rosettes. Large, barrel-shaped, upright cones with sunken tops. Large, spreading tree with branches rising upwards.

25m

Top not sunken

Cones are covered with sticky resin

▲ Cedar of Lebanon

Similar to Atlas Cedar, but cones a little larger, tops not sunken. Branches level, lower ones carrying table-like masses of foliage.

30m

Leaves overlap each other

Twisting branches

23m

▲ Chile Pine

Also called Monkey Puzzle. Stiff, leathery, triangular leaves with sharp points growing all round the shoot. Broad round crown. Pole-like trunk with wrinkled bark.

Drooping top shoot and branch tips

23m

▲ Deodar

Leaves like other Cedars, but longer, softer, and paler green. Large, barrel-shaped cones have sunken top. Tall tree with pointed crown.

Long-stalked, tall acorn

Acorn cup

Lobe

Long stalk

23m

▲ English Oak

Leaves short-stalked with ear-like lobes at base. Broad crown. Trunk shorter than Sessile Oak. Many large branches growing from same point.

21m

▲ Sessile Oak

Thick, dark green, long-stalked leaves tapering to base. Branches grow from stem at different levels and point upwards in narrow crown.

All veins go to tips of lobes

Acorn more rounded than on Common Oak

Often stalkless

Leaf shape varies

Teeth

Evergreen leaves

Small acorn, almost covered by cup

20m

▲ Holm Oak
Shiny evergreen leaves, greyish-green beneath, sometimes with shallow teeth like Holly leaves. Common ornamental tree. Broad dense crown.

25m

▲ Turkey Oak
Leaves unevenly lobed and saw-toothed. Whiskers on buds and at base of leaves. Acorn cups mossy and stalkless. Acorns ripen in second autumn.

Acorn cup is mossy

16m

Twisted trunk and branches

▲ Cork Oak

Shiny evergreen leaves with wavy edge. Thick, corky, whitish bark. Small tree, except in warmer countries. Common in southern Europe.

Acorn

Cork is obtained from the bark

Autumn colour of leaves

20m

▲ Red Oak

Large leaves with bristly-tipped lobes, turn reddish-brown in autumn. Squat acorns in shallow cups ripen in second year. Smooth silvery bark.

Acorn

25m

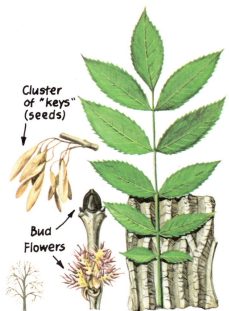

Cluster
of "keys"
(seeds)

Bud
Flowers

▲ Common Ash

Compound leaf of 9-13
leaflets appearing late,
after bunches of purplish
flowers. Clusters of "keys"
stay on the tree
into winter. Pale
grey bark.

20m

Fruit

Flowers

Leaflets
downy
near veins

▲ Manna Ash

Compound leaf of 5-9
stalked leaflets. Clusters
of showy white flowers
in May. Smooth grey bark
oozes sugary
liquid called
manna.

Leaf has notched tip

▲ Common Alder
Leaves rounded, falling in late autumn. Young twigs and leaves sticky. Reddish catkins. Fruits like small, brown, woody cones. Always near water.

Cone-like fruit stays on all winter

Green fruits ripen into brown, woody "cones" (shown above)

12m

14m

▲ Grey Alder
Pointed oval leaves with sharply-toothed edge, soft and grey underneath. Twigs and bark grey. Catkins and fruit like Common Alder.

Berries

One flower (from a cluster)

Toothed edge

Leaves turn red in Autumn

7m

▲ Rowan

Compound leaf like Ash, but smaller. Clusters of creamy-white flowers in May. Red berries ripen in August. Small tree. Often grows alone on mountainsides.

8m

▲ Whitebeam

Large oval leaves with toothed edge, white and furry underneath. Flowers and fruit like Rowan but ripen later. Grows at edges of woods.

Berries

Leaf stalk is flattened →

20m

▲ Aspen

Rounded leaves with wavy edge, trembling in wind. White downy catkins. Grey bark with large pores. Smaller than other Poplars. Often in thickets.

Fan-shaped crown

25m

▲ Black Italian Poplar

Dark green, triangular, pointed leaves, appearing late. Red catkins. Deeply furrowed bark. Trunk and crown often lean away from wind. Grows fast.

Wavy edges

Under-side of leaf

20m

▲ White Poplar

Five-lobed leaves are downy white underneath. Entire crown looks white. Lower bark dark and rugged, upper bark pale grey. Crown often leaning.

Lower leaves are less lobed

Diamond-shaped marks on young bark

Large, fast-growing tree

35m

▲ Western Balsam Poplar

Large, triangular, pointed leaves, white underneath. Buds and young leaves sticky and sweet-smelling. Long purplish catkins. White fluffy seeds.

Underside of leaf

Leaf shape varies

28m

▲ Lombardy Poplar
Pointed triangular leaves.
Tall narrow tree. Branches
grow upwards from
ground. Furrowed
bark. Often along
roadsides.

High-domed
crown

Leaf from
upper
branch

Rounded
leaf from
lower branch

23m

▲ Grey Poplar
Similar to White Poplar.
Wavy-edged leaves, never
deeply lobed, downy white
underneath. Upper bark
yellowish-grey,
lower bark
dark, furrowed.

Catkin
(Pussy
Willow)

Underside
of leaf

▲ Goat Willow

Broad, rounded, rough,
grey-green leaves. Silvery-
grey, upright catkins in late
winter. Small bushy tree.
Common on damp
waste ground
and scrub
woodland.

7m

The crown shape
varies

Broad crown

15m

▲ Crack Willow

Very long, narrow leaves,
bright green above, grey-
green below. Twigs snap
easily. Grows near water,
often with
branches cut
back to trunk.

Underside
of leaf

Catkin

20m

▲ **White Willow**

Long, narrow, finely-toothed leaves, white underneath. Slender twigs, hard to break. Common by water. Weeping Willow is a variety with trailing branches.

Underside of leaf →

Catkin ↓

15m

▲ **Silver Birch**

Small, diamond-shaped leaves with double-toothed edge. Long "lamb's tail" catkins in April. Slender tree with drooping branches.

Catkin

Silvery bark peels off in ribbons →

Leaves are wavy-edged

Nuts in husk

▲ Common Beech

Light green, oval leaves turn copper-brown in autumn. Triangular nuts in hairy husks. Tall tree with spreading crown. Smooth grey bark.

25m

▲ Hornbeam

Sharply-toothed, oval leaves. In autumn, clusters of three-pronged, leaf-like wings hold nuts. Smooth grey bark is fluted (or rippled).

10m

Cluster of green winged fruits

10m

▲ **Crab Apple**

Small rounded leaves with toothed edge. Pinkish-white flowers in May. Small, speckled, reddish-green apples. Small bushy tree. Common in hedges.

Apple tastes sour

15 m

▲ **Common Pear**

Small, dark green, oval leaves with long stalks. Large, showy, white flowers in April. Small pears are gritty to eat. Tall narrow tree. In woods and hedgerows.

Pear is golden when ripe

20m

▲ Southern Beech

Narrow oval leaves with fine-toothed edge and many obvious veins. Deep green, prickly fruit. Silver-grey bark. Triangular-shaped crown.

Fruits

Leaves on short stalks

Leaves are hairy underneath

(Rare in Britain)

20m

▲ European White Elm

Large, rough, oval leaves, lop-sided at base, with double-toothed edge. Flowers and fruits on long stalks. Tall tree with broad open crown.

30m

▲ London Plane

Large broad leaves with
pointed lobes. Spiny
"bobble" fruits hanging
all winter. Flaking bark
leaving yellowish
patches. Tall tree,
often in towns.

Fruit

20m

▲ Sycamore

Dark green, leathery leaves
with five lobes. Paired,
closely-angled, winged
seeds. Large spreading
tree. Smooth
brown bark
becoming scaly.

Toothed
edge

Seeds twist
as they fall

Leaves turn golden in autumn

Pairs of seeds spin as they fall

Lobes are blunt

Leaves turn golden in autumn

Seeds

15m

▲ Norway Maple

Light green, thin leaves.
Lobes and teeth are
bristle-tipped. Paired
seeds form wide angle.
Smaller, less spreading
than Sycamore.
Finely-furrowed,
grey bark.

10m

▲ Field Maple

Small, dark green leaves
with five lobes. Small,
reddish, winged seeds
form a straight line. Small
tree with round
head. Often in
hedges.

▲ Common Lime

Broad crown

Leafy wing

Fruits

Pointed tip

25m

Heart-shaped leaves with toothed edge. Yellowish-green, sweet-smelling flowers in July. Small, round, hard, grey-green fruits hang from leafy wing.

Rounded crown

Leafy wing

Fruits

20m

▲ Silver Lime

Very similar to Common Lime, but leaves dark green above, silvery-grey and hairy underneath. Crown more rounded than Common Lime.

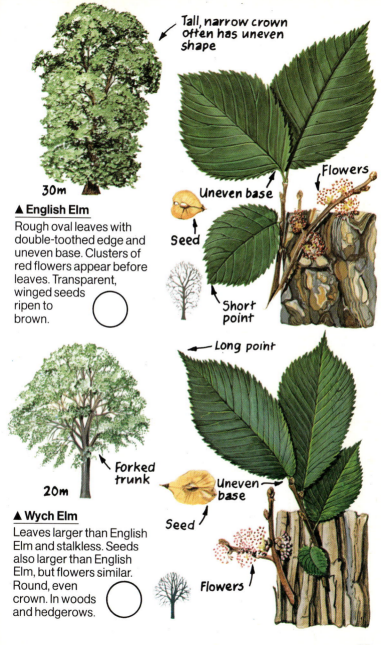

Tall, narrow crown often has uneven shape

Flowers

Uneven base

Seed

Short point

30m

▲ English Elm

Rough oval leaves with double-toothed edge and uneven base. Clusters of red flowers appear before leaves. Transparent, winged seeds ripen to brown.

Long point

Uneven base

Seed

Forked trunk

Flowers

20m

▲ Wych Elm

Leaves larger than English Elm and stalkless. Seeds also larger than English Elm, but flowers similar. Round, even crown. In woods and hedgerows.

Tree
in bloom

"Candle"
of flowers

Leaflet

25m

▲ Horse Chestnut
Compond leaf made of 5-7
large leaflets. Upright
"candle" of white (or pink)
flowers in May. Brown
"conker" in green
spiny case. Parks
and avenues.

Conker
(fruit)

Flowers

Clusters
of 2-3
fruits
containing
nuts

25m

▲ Sweet Chestnut
Long narrow leaves with
saw-toothed edge. Edible
brown chestnuts in green
prickly case. Spiral-
furrowed bark.
Large, tall-
crowned tree.

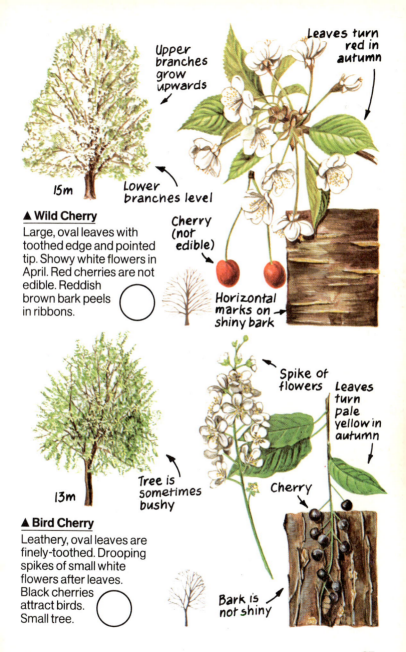

Upper branches grow upwards

Leaves turn red in autumn

15m

Lower branches level

Cherry (not edible)

Horizontal marks on shiny bark

▲ Wild Cherry

Large, oval leaves with toothed edge and pointed tip. Showy white flowers in April. Red cherries are not edible. Reddish brown bark peels in ribbons.

Spike of flowers

Leaves turn pale yellow in autumn

13m

Tree is sometimes bushy

Cherry

Bark is not shiny

▲ Bird Cherry

Leathery, oval leaves are finely-toothed. Drooping spikes of small white flowers after leaves. Black cherries attract birds. Small tree.

Unripe fruit

Ripe fruit

Young fruits

▲ Black Mulberry

Rough, heart-shaped leaves with toothed edge. Short catkins. Edible, blackish-red berries. Low, broad-crowned tree. Short trunk and twisted branches.

12m

Old trees have branches to the ground and often lean over

Smooth, green case containing edible walnut

Young fruit

▲ Common Walnut

Compound leaves of 7-9 untoothed leaflets. Twigs are hollow, with cross-sections inside. Smooth grey bark with some cracks, or fissures. Broad crown.

15m

Leaves are bronze when they first open, turning green later

Very sharp thorns ⟶

Smooth-edged leaflet

20m

▲ False Acacia

Compound leaves of many small leaflets. Pairs of sharp thorns on twigs. Hanging clusters of white flowers in June. Seeds in pods. Deeply-furrowed bark.

Tree often has several trunks

Tree in bloom (May–June)

Young seed-pods are green

Leaflets are soft and hairy

7m

▲ Laburnum

Leaf made up of three leaflets. Hanging clusters of yellow flowers. Poisonous seeds in twisted brown pods. Small tree. Smooth, green-brown bark.

Leaves are thick
and leathery

Berries
appear only
on the female
trees

10m

▲ **Holly**
Shiny, dark, evergreen
leaves with thorny prickles.
Small white flowers. Round
red berries. Smooth, grey-
green bark.
Small tree
or shrub.

Two kinds
of flower

Flowers

3m
▲ **Tamarisk**   Tree in bloom
Tiny, grey-green, scale-
like leaves, which look
feathery. Clusters of small
pinkish-white flowers.
Shrub or small tree with
slender
branches. Often
near the sea.

Leaves

Twig

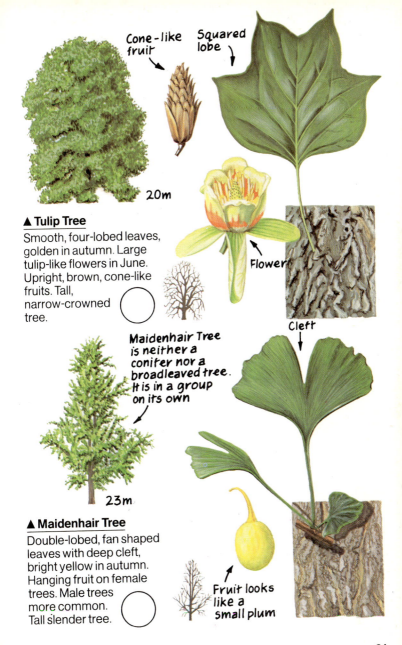

Cone-like fruit

Squared lobe

20m

Flower

Cleft

▲ Tulip Tree

Smooth, four-lobed leaves, golden in autumn. Large tulip-like flowers in June. Upright, brown, cone-like fruits. Tall, narrow-crowned tree.

Maidenhair Tree is neither a conifer nor a broadleaved tree. It is in a group on its own

23m

▲ Maidenhair Tree

Double-lobed, fan shaped leaves with deep cleft, bright yellow in autumn. Hanging fruit on female trees. Male trees more common. Tall slender tree.

Fruit looks like a small plum

CHAPTER 3

Fungi

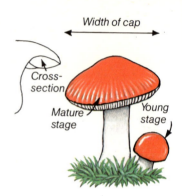

Width of cap

Cross-section

Mature stage

Young stage

This Chapter will help you to identify many of the mushrooms, toadstools and other fungi that grow in Britain and the rest of Europe.

Each illustration shows the mature (fully grown) stage of a fungus and very often a young stage and a cross-section are also shown.

Beside each illustration is a description which tells you where it grows and gives its size.

Some fungi are quite difficult to identify and may not look exactly like their picture, so it is very important to read the description and notes around the picture.

Measurements
Measurements given for mushrooms, toadstools and bracket fungi are for the width of the cap.

For other fungi without a cap, measurements are for the height or width of the whole fungus.

As each kind of fungus varies in size, depending on the conditions that it grows in, a maximum and minimum size is given for the mature stage.

WARNING
Quite a few fungi are **deadly poisonous.** Some edible and harmless fungi look very similar to those which are poisonous. So **never eat or taste** any fungus, unless an expert has helped you to identify it.

Where to look
Each type of fungus grows in a particular place or habitat. Some grow only in broadleaved woods, such as beechwoods, others grow only in coniferous woods under trees such as pine. Some fungi grow in open grassland. Wherever there are trees, dead wood or grass and moist conditions, there are usually some fungi to be found.

In each habitat there will be different fungi growing on the ground, on tree trunks, on dead wood and in leaf litter. Look carefully in all of these places.

When to look
Although most fungi appear in the autumn, some interesting species (or kinds) can be found at other times of year. The descriptions next to each illustration tell you when each species can be seen.

Identifying mushrooms and toadstools

In this Chapter only the cap fungi with gills belonging to the *Agaricus* group (p.104-5) are called mushrooms. The word "toadstool" is used for all other cap fungi, whether they have gills, pores, or spines. This page tells you what to note when you are trying to identify a mushroom or a toadstool.

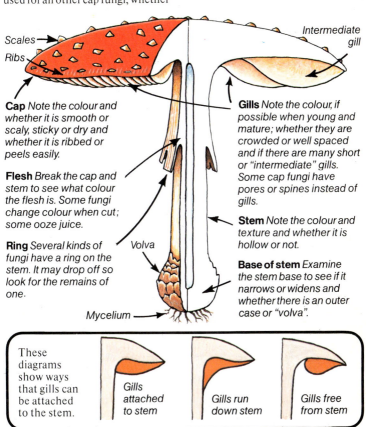

Scales

Ribs

Intermediate gill

Cap Note the colour and whether it is smooth or scaly, sticky or dry and whether it is ribbed or peels easily.

Flesh Break the cap and stem to see what colour the flesh is. Some fungi change colour when cut; some ooze juice.

Ring Several kinds of fungi have a ring on the stem. It may drop off so look for the remains of one.

Volva

Mycelium

Gills Note the colour, if possible when young and mature; whether they are crowded or well spaced and if there are many short or "intermediate" gills. Some cap fungi have pores or spines instead of gills.

Stem Note the colour and texture and whether it is hollow or not.

Base of stem Examine the stem base to see if it narrows or widens and whether there is an outer case or "volva".

These diagrams show ways that gills can be attached to the stem.

Gills attached to stem

Gills run down stem

Gills free from stem

Other points to note

Smell – Some fungi have a distinct smell that helps to identify them. **How and where it grows** – Note whether the fungus grows singly, in clusters or in a ring. Note where it grows and which trees are nearby. **When** – Note the time of year.

Scientific names

The scientific name of each fungus is given under its English name. Some species have only a scientific name. The first word of the scientific name is the group that the fungus belongs to. The second word gives the name of the species of fungus.

93

Different types of fungi

Fungi are "plants" that cannot make their own food and so have to live off other plants or animals. There are many thousands of different fungi, ranging in size and shape from tiny moulds to large bracket fungi. This Chapter shows only some of the larger ones that belong to the groups illustrated below. To find out more about them, turn to the page listed beside each picture below.

All the illustrations in this Chapter show the fruit bodies of fungi. Each fruit body produces thousands of tiny dust-like spores from which new fungi can grow (see opposite). The main part of the fungus exists as a mass of tiny threads, known as "mycelium". Fungus mycelium lives all year round buried in the plant or animal matter that it feeds on. If you pick a mushroom or toadstool, you will see some of the mycelium attached to the base of the stem.

The groups of fungi illustrated in this Chapter include:

Cap fungi with pores (p. 96)
Pores

**1 Morels
2 Stinkhorn** (p. 132-3)

Cap fungi with gills (p. 100)
Gills

**1 Cup fungi
2 Bird's Nest Fungus
3 Truffle** (p. 134-5)

**1 Puffballs
2 Earthball
3 Earth Star** (p. 136-7)

Cap fungi with spines (p. 126)
Spines

Jelly fungi (p. 138)

Bracket fungi (p. 127)

Crust fungi (p. 131)

Fairy Clubs (p. 139)

How a fungus grows

Both the mycelium, which lives buried in the soil, and the mushroom (or fruit body), which appears above ground, are made up of tiny thread-like tubes called "hyphae". The mycelium is made up of loosely arranged hyphae and the mushroom is made up of tightly packed hyphae. Hyphae develop from spores that are produced in the gills of a mushroom. The illustrations below show how a mushroom develops and how it completes the life cycle of a fungus.

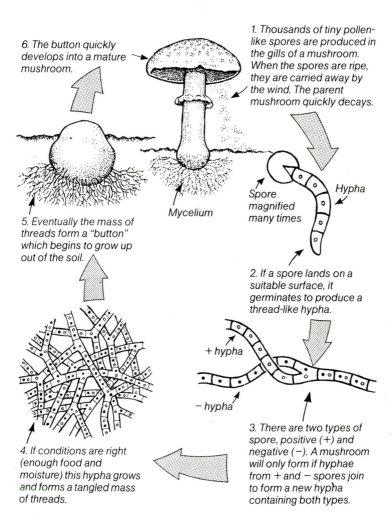

6. The button quickly develops into a mature mushroom.

1. Thousands of tiny pollen-like spores are produced in the gills of a mushroom. When the spores are ripe, they are carried away by the wind. The parent mushroom quickly decays.

Mycelium

Spore magnified many times

Hypha

5. Eventually the mass of threads form a "button" which begins to grow up out of the soil.

2. If a spore lands on a suitable surface, it germinates to produce a thread-like hypha.

+ hypha

− hypha

4. If conditions are right (enough food and moisture) this hypha grows and forms a tangled mass of threads.

3. There are two types of spore, positive (+) and negative (−). A mushroom will only form if hyphae from + and − spores join to form a new hypha containing both types.

Cap fungi with pores

The cap fungi on pages 96-9 all have pores underneath the cap. These pores are the openings of a mass of spongy tubes where the spores are produced. The flesh and pores of many species change colour when cut or bruised.

White flesh does not change colour when cut

Pores come away easily from cap

Swollen stem with raised veins

Pale rim round edge of cap

Penny Bun ▶
Boletus edulis
Light to dark brown cap, feels greasy when wet. Pores at first white, then pale yellow when mature. Grows in broadleaved woods.
Cap 5-18 cm.
Sept.-Nov.

Cap often flat when mature

Flesh turns blue when cut

Cep ▶
Boletus badius
Dark brown cap feels velvety when dry, sticky when wet. White pores when young. Usually under conifers.
Cap 5-12 cm.
Aug.-Nov.

Yellow pores turn blue when bruised

Cap fungi with pores

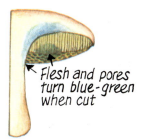

Flesh and pores turn blue-green when cut

Red-cracked Boletus ▶

Boletus chrysenteron
When mature, red-brown cap flattens and cracks to show red flesh. Grows in broadleaved woods.
Cap 4-12 cm.
Aug.-Nov.

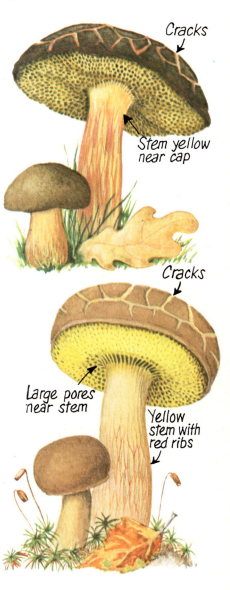

Cracks

Stem yellow near cap

Cracks

Large pores near stem

Yellow stem with red ribs

Yellow flesh and pores

Yellow-cracked Boletus ▶

Boletus subtomentosus
Light to olive brown cap feels velvety and when mature, often cracks to show yellow flesh. Broadleaved woods.
Cap 4-12 cm.
Aug.-Nov.

Cap fungi with pores

Do not taste any fungus without expert advice.

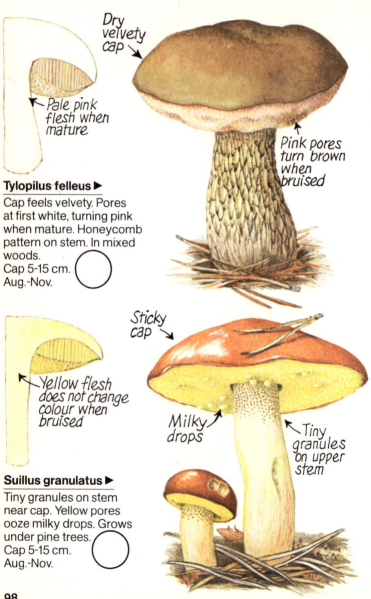

Dry velvety cap →

← Pale pink flesh when mature

Pink pores turn brown when bruised

Tylopilus felleus ▶
Cap feels velvety. Pores at first white, turning pink when mature. Honeycomb pattern on stem. In mixed woods.
Cap 5-15 cm.
Aug.-Nov.

Sticky cap →

Yellow flesh does not change colour when bruised

Milky drops →

← Tiny granules on upper stem

Suillus granulatus ▶
Tiny granules on stem near cap. Yellow pores ooze milky drops. Grows under pine trees.
Cap 5-15 cm.
Aug.-Nov.

Cap fungi with pores

Dirty white pores darken when bruised

Tall stem with tufts of darker scales

White flesh turns pink when cut

◄ Rough Stemmed Boletus

Leccinum scabrum
Stem covered with tufts of scales. Pores off-white. Under birch trees. Cap 6-20 cm. Aug.-Nov.

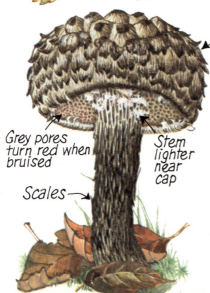

Dark woolly scales on cap

Flesh turns red then black when bruised

Grey pores turn red when bruised

Stem lighter near cap

Scales

◄ Old Man of the Woods

Strobilomyces floccopus
Cap covered with scales, grey at first then brown-black when mature. Dries out without rotting. Broad-leaved woods. Cap 8-15 cm. Sept.-Nov.

Cap fungi with gills

The **Amanitas**, pages 100-2, develop from an "egg", enclosed in a white veil, which splits as the stem grows. The remains of the veil at the base of the stem form a "volva". Remains of this veil may also form warts on the cap. Later, a second veil, covering the gills, splits and forms a ring around the stem, although some species do not have a ring. Amanitas have white gills that are free from the stem, and white spores. Several are very poisonous.

Gills free from stem

Ribs on cap edge

Warts on cap sometimes missing

Rings on volva

Panther Cap ▶

Amanita pantherina
Large ring. White gills. Grows in clearings in broadleaved woods, usually near beech trees. Deadly poisonous. Cap 6-12 cm. Aug.-Oct.

White flesh turns slightly red when cut

White gills

Large ribbed ring

Warty scales on volva

Blusher ▶

Amanita rubescens
Flesh is tinged red when damaged. In broadleaved and coniferous woods. Cap 5-15 cm. July-Oct.

Cap fungi with gills

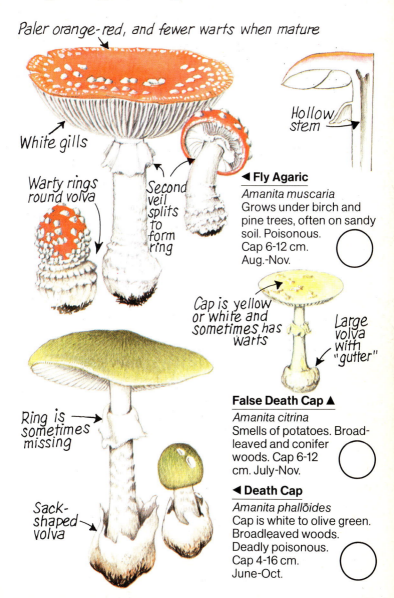

Paler orange-red, and fewer warts when mature

White gills

Warty rings round volva

Second veil splits to form ring

Hollow stem

◀ Fly Agaric

Amanita muscaria
Grows under birch and pine trees, often on sandy soil. Poisonous.
Cap 6-12 cm.
Aug.-Nov.

Cap is yellow or white and sometimes has warts

Large volva with "gutter"

False Death Cap ▲

Amanita citrina
Smells of potatoes. Broad-leaved and conifer woods. Cap 6-12 cm. July-Nov.

Ring is sometimes missing

◀ Death Cap

Amanita phallōides
Cap is white to olive green.
Broadleaved woods.
Deadly poisonous.
Cap 4-16 cm.
June-Oct.

Sack-shaped volva

Cap fungi with gills

White gills

Tall, hollow stem

Ragged scale markings on stem

Sack-shaped volva

Destroying Angel ▶
Amanita virosa
Cap at first conical,
expanding with age. In
conifer and broadleaved
woods. Rare. Deadly
poisonous.
Cap 6-12 cm.
July-Oct.

Ribs on edge of cup

Hollow stem without ring

Stem paler than cap

Tawny Grisette ▶
Amanita fulva
Grows in broadleaved
woods especially under
birch on poor soil.
Cap 3-10 cm.
May-Nov.

Sack-shaped volva

Cap fungi with gills

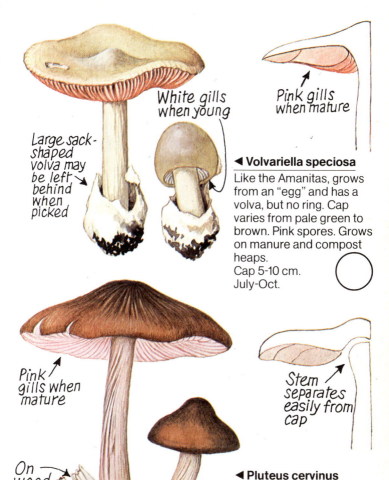

White gills when young

Pink gills when mature

Large sack-shaped volva may be left behind when picked

◀ Volvariella speciosa

Like the Amanitas, grows from an "egg" and has a volva, but no ring. Cap varies from pale green to brown. Pink spores. Grows on manure and compost heaps.
Cap 5-10 cm.
July-Oct.

Pink gills when mature

Stem separates easily from cap

On wood

◀ Pluteus cervinus

On sawdust and stumps of broadleaved trees. Pink spores. Cap 3-12 cm. Grows all year round but especially May-Nov.

Cap fungi with gills

Mushrooms (*Agaricus* group), pages 104-5, develop from a "button" that is covered in a veil that breaks to leave a ring round the stem. The gills are at first pink or grey, but never white, turning dark brown when mature, and are free from the stem. Mushrooms have chocolate-brown spores. Not all are edible.

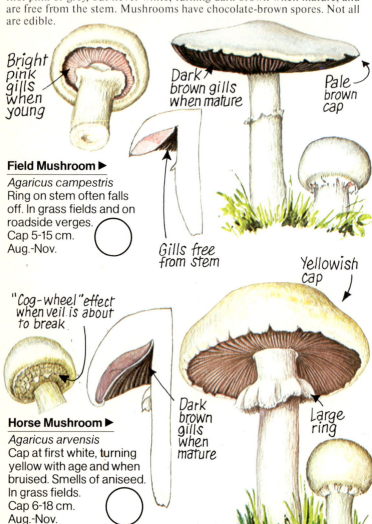

Bright pink gills when young

Dark brown gills when mature

Pale brown cap

Field Mushroom ▶

Agaricus campestris
Ring on stem often falls off. In grass fields and on roadside verges.
Cap 5-15 cm.
Aug.-Nov.

Gills free from stem

Yellowish cap

"Cog-wheel" effect when veil is about to break

Horse Mushroom ▶

Agaricus arvensis
Cap at first white, turning yellow with age and when bruised. Smells of aniseed. In grass fields.
Cap 6-18 cm.
Aug.-Nov.

Dark brown gills when mature

Large ring

Cap fungi with gills

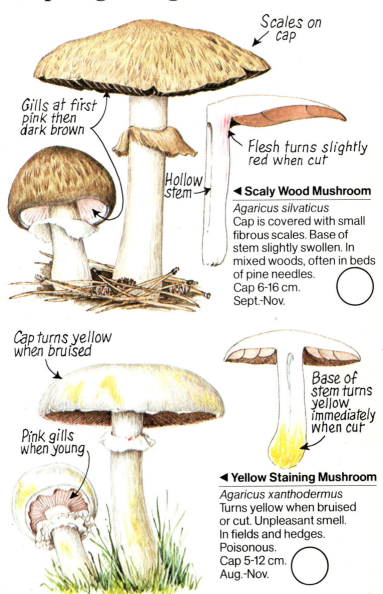

Scales on cap

Gills at first pink then dark brown

Flesh turns slightly red when cut

Hollow stem

◄ Scaly Wood Mushroom

Agaricus silvaticus
Cap is covered with small fibrous scales. Base of stem slightly swollen. In mixed woods, often in beds of pine needles.
Cap 6-16 cm.
Sept.-Nov.

Cap turns yellow when bruised

Base of stem turns yellow immediately when cut

Pink gills when young

◄ Yellow Staining Mushroom

Agaricus xanthodermus
Turns yellow when bruised or cut. Unpleasant smell. In fields and hedges. Poisonous.
Cap 5-12 cm.
Aug.-Nov.

Cap fungi with gills

Parasols, page 106, have whitish gills that are free from the stem, and white spores. They have a ring, but no volva, and the stem separates easily from the cap.

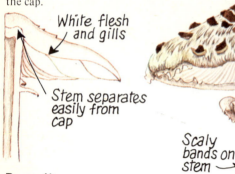

White flesh and gills

Scales

Stem separates easily from cap

Ring

Scaly bands on stem

Cap almost smooth when young

Parasol ▶
Lepiota procera
Stem has snake-like patterns and is swollen at base. In woods and grassy places.
Cap 5-15 cm.
July-Nov.

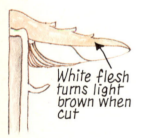

White flesh turns light brown when cut

Scales

Ring

Smooth stem turns red when bruised

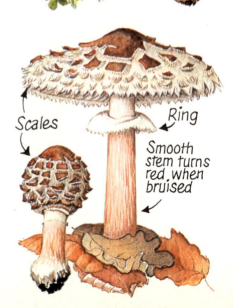

Shaggy Parasol ▶
Lepiota rhacodes
Smooth stem, swollen at base. In clearings in woods and grassy places.
Cap 5-15 cm.
July-Nov.

Cap fungi with gills

Ink Caps, pages 107-8, have thin crowded gills that often dissolve into a black inky liquid with age. The spores are black.

Cap dissolves with age

Gills white at first, then pink and finally black

White scales

Ring

◄ Shaggy Ink Cap

Coprinus comatus
Slender, hollow stem separates easily from the cap. Often in groups in fields and on roadside verges.
Cap 5-10 cm high.
May-Nov.

Cap dissolves with age

White veil covers cap when young

◄ Magpie Ink Cap

Coprinus picaceus
Brown-black cap with white patches. Gills at first white, then pale brown, finally black. Broadleaved woods.
Cap 5-10 cm high.
Sept.-Nov.

Cap fungi with gills

Common Ink Cap ▶

Coprinus atramentarius
Gills at first dirty white, turning brown then black. Cap dissolves with age. Usually grows in groups at base of trees, in fields or in woods.
Cap 3-7 cm.
May-Dec.

Ring-like zones at base of stem

Ribs on edge of cap

Ribs on cap

Granules

◀ Glittering Ink Cap

Coprinus micaceus
Tiny granules on cap when young. Gills at first white then black, only dissolving slightly. Grows in groups on tree stumps.
Cap 2-5 cm high.
May-Dec.

Snowy Ink Cap ▶

Coprinus niveus
Gills at first grey, then black. Cap curls and dissolves with age. On cow or horse dung.
Cap 1-3 cm.
May-Nov.

Cap fungi with gills

Ribs on edge of cap

Purple brown gills

One "root"

Bolbitius vitellinus ▶
Cinnamon coloured spores. On grass and straw. Cap 2-5 cm. July-Oct.

Small ring is sometimes torn or missing

◀ Psathyrella multipedata
Grows in groups of ten or more from a common "root". Chocolate-brown spores. By paths in woods. Cap 2-3 cm. July-Oct.

Bright yellow cap with ribbed edge

Rust coloured gills

Mottled grey-black gills

◀ Dung Roundhead
Paneolus semiovatus. Bell-shaped cap feels sticky when wet. Black spores. On dung. Cap 2-5 cm. July-Nov.

Cap fungi with gills

Yellow flesh

Dark brown gills when old

Orange tint at cap centre

Yellow-green gills

Stem darker at base

Sulphur Tuft ▶

Hypholoma fasciculare
Faint ring on stem. Yellow flesh. Grows in clusters on broadleaved tree stumps, often in large numbers. Purple-brown spores. Cap 4-10 cm. Aug.-Nov.

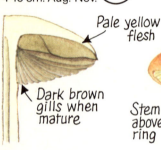

Pale yellow flesh

Dark brown gills when mature

Stem smooth above faint ring

Brick-red Cap ▶

Hypholoma sublaterium
Gills at first yellow, then lilac-grey, finally dark brown. Purple-brown spores. Grows in clusters on tree stumps. Cap 3-8 cm. Sept.-Dec.

110

Cap fungi with gills

Sticky cap

Dark velvety stem covered with tiny hairs

Pale yellow gills attached to stem

◀ Velvet Stem

Flammulina velutipes
Stem ends in a root-like thread. Gills turn brown with age. On trunks, stumps and branches of broad-leaved trees.
Cap 3-8 cm.
Sept.-March.

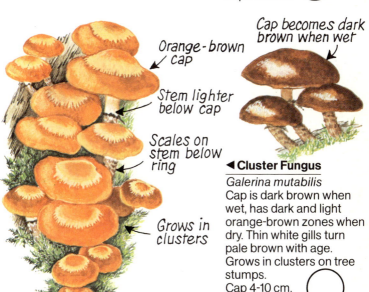

Orange-brown cap

Stem lighter below cap

Scales on stem below ring

Grows in clusters

Cap becomes dark brown when wet

◀ Cluster Fungus

Galerina mutabilis
Cap is dark brown when wet, has dark and light orange-brown zones when dry. Thin white gills turn pale brown with age. Grows in clusters on tree stumps.
Cap 4-10 cm.
All year round.

Cap fungi with gills

Brown gills when mature

Scales

Stem smooth above ring

Shaggy Pholiota ▶

Pholiota squarrosa
Yellow gills turn brown with age. Grows in groups at base of broadleaved trees.
Cap 3-8 cm.
Sept.-Nov.

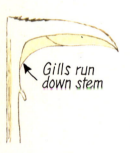

Gills run down stem

Honey Fungus ▶

Armillaria mellea
Cream coloured gills turn brown with age. Grows at base of broadleaved or conifer trees which it eventually kills.
Cap 3-10 cm.
July-Dec.

Scales on cap

Ring →

Black "boot-laces" between bark and wood

Cap fungi with gills

Wax Caps, page 113, are mainly white or brightly coloured toadstools which look waxy and have well-spaced gills. They have white spores.

Yellow gills when young

Base of stem is white

Widely spaced red gills

◄ Crimson Wax Cap

Hygrophorus puniceus
Cap colour fades with age. Outline of gills can be seen through the thin cap when it is held up to the light.
In grass fields and roadsides.
Cap 5-12 cm.
Aug.-Dec.

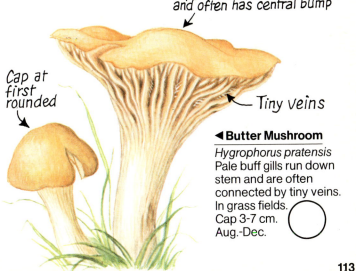

Cap flattens when mature and often has central bump

Cap at first rounded

Tiny veins

◄ Butter Mushroom

Hygrophorus pratensis
Pale buff gills run down stem and are often connected by tiny veins.
In grass fields.
Cap 3-7 cm.
Aug.-Dec.

Cap fungi with gills

Clitocybes, pages 114-5, have whitish or pale gills that run down the stem and the cap is often funnel-shaped. They have white spores.

Gills run down stem

Base of stem is swollen and woolly

Aniseed Toadstool ▶
Clitocybe odora
Blue-green cap with central bump. Amongst dead leaves in broad-leaved woods. Smells of aniseed.
Cap 3-7 cm.
Aug.-Nov.

Flesh is thick at centre

Cap darker grey at centre

Edge of cap curves under

Cloudy Cap ▶
Clitocybe nebularis
Grey to grey-brown cap. Grows in woods, especially under pine. Slightly sickly smell.
Cap 6-25 cm.
Aug.-Nov.

Cap fungi with gills

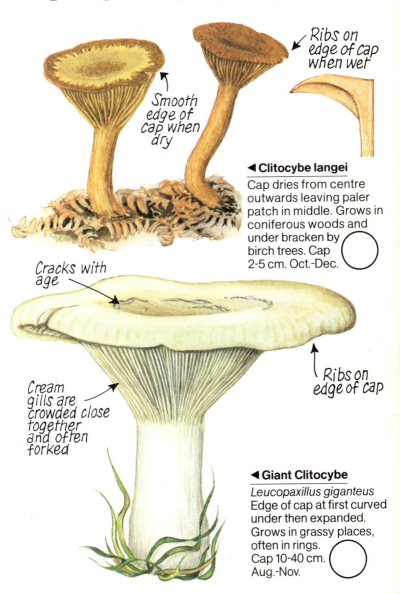

Ribs on edge of cap when wet

Smooth edge of cap when dry

◀ Clitocybe langei
Cap dries from centre outwards leaving paler patch in middle. Grows in coniferous woods and under bracken by birch trees. Cap 2-5 cm. Oct.-Dec.

Cracks with age

Ribs on edge of cap

Cream gills are crowded close together and often forked

◀ Giant Clitocybe
Leucopaxillus giganteus
Edge of cap at first curved under then expanded. Grows in grassy places, often in rings. Cap 10-40 cm. Aug.-Nov.

Cap fungi with gills

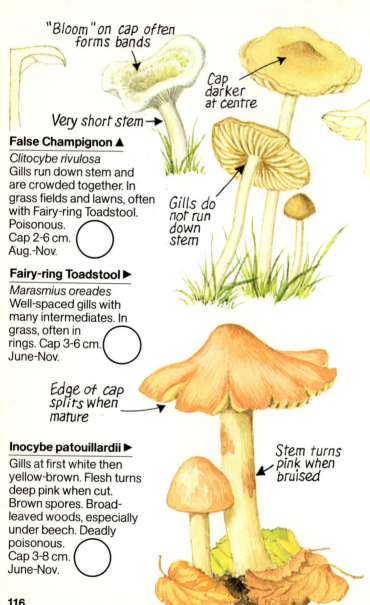

"Bloom" on cap often forms bands

Cap darker at centre

Very short stem →

Gills do not run down stem

False Champignon ▲
Clitocybe rivulosa
Gills run down stem and are crowded together. In grass fields and lawns, often with Fairy-ring Toadstool. Poisonous. Cap 2-6 cm. Aug.-Nov.

Fairy-ring Toadstool ▶
Marasmius oreades
Well-spaced gills with many intermediates. In grass, often in rings. Cap 3-6 cm. June-Nov.

Edge of cap splits when mature

Inocybe patouillardii ▶
Gills at first white then yellow-brown. Flesh turns deep pink when cut. Brown spores. Broad-leaved woods, especially under beech. Deadly poisonous. Cap 3-8 cm. June-Nov.

Stem turns pink when bruised

Cap fungi with gills

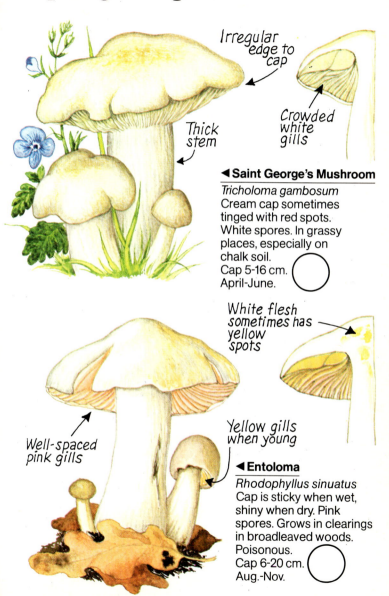

Irregular edge to cap

Crowded white gills

Thick stem

◀ Saint George's Mushroom

Tricholoma gambosum
Cream cap sometimes
tinged with red spots.
White spores. In grassy
places, especially on
chalk soil.
Cap 5-16 cm.
April-June.

White flesh
sometimes has
yellow
spots

Well-spaced
pink gills

Yellow gills
when young

◀ Entoloma

Rhodophyllus sinuatus
Cap is sticky when wet,
shiny when dry. Pink
spores. Grows in clearings
in broadleaved woods.
Poisonous.
Cap 6-20 cm.
Aug.-Nov.

Cap fungi with gills

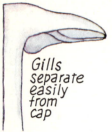

Gills separate easily from cap

Wood Blewit ▶

Lepista nuda
Tinged lilac all over. Cap colour fades with age. Gills crowded close together. Pale pink spores. Grows in broadleaved and coniferous woods. Cap 6-15 cm. Sept.-Feb.

White downy base of stem

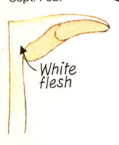

White flesh

Field Blewit ▶

Lepista saeva
Cap varies from pale to dark grey-brown. Cream gills are crowded together. Pale pink spores. In grass fields, wood clearings and hedges. Cap 5-15 cm. Sept.-Dec.

Stem tinged lilac

Cap fungi with gills

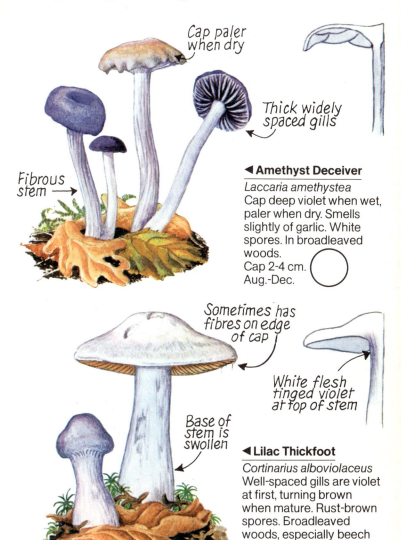

Cap paler when dry

Thick widely spaced gills

Fibrous stem →

◀ Amethyst Deceiver

Laccaria amethystea
Cap deep violet when wet, paler when dry. Smells slightly of garlic. White spores. In broadleaved woods.
Cap 2-4 cm.
Aug.-Dec.

Sometimes has fibres on edge of cap

White flesh tinged violet at top of stem

Base of stem is swollen

◀ Lilac Thickfoot

Cortinarius alboviolaceus
Well-spaced gills are violet at first, turning brown when mature. Rust-brown spores. Broadleaved woods, especially beech and oak.
Cap 3-8 cm.
Aug.-Dec.

Cap fungi with gills

Russulas, pages 120-1, have crumbly flesh. The spores and gills are white or cream. Most gills run right from the edge of the cap to the stem, with very few short or "intermediate" gills in between. Many Russulas are brightly coloured. There are almost 100 species in Britain.

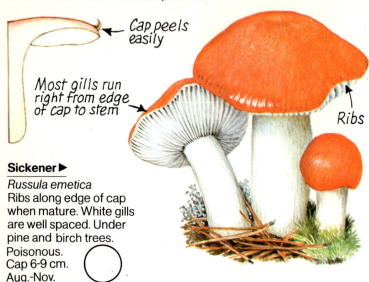

Cap peels easily

Most gills run right from edge of cap to stem

Ribs

Sickener ▶

Russula emetica
Ribs along edge of cap when mature. White gills are well spaced. Under pine and birch trees. Poisonous. Cap 6-9 cm. Aug.-Nov.

Gills sometimes have fine rust-coloured spots on edges

Thick stem →

Well-spaced cream gills

Russula atropurpurea ▶

Centre of cap is darker colour and often depressed. Broadleaved woods. Cap 5-15 cm. July-Nov.

Cap fungi with gills

Ribs on edge of cap

White stem with faint veins

◄ **Ochre Russula**

Russula ochroleuca
White to pale cream gills break easily. White flesh turns light grey with age. All types of wood, but especially under pine. Cap 4-10 cm. Aug.-Nov.

Scales

Green Russula ►

Russula virescens
Green cap with darker scales. White gills break easily and are crowded. Broadleaved woods, especially beech. Cap 5-12 cm. July-Oct.

Cap cracks with age

Stem narrower at base

Cream gills feel greasy and are elastic

◄ **Variable Russula**

Russula cyanoxantha
Cap varies from mottled green to violet or grey. Flesh is brown when cut. Broadleaved woods, especially beech. Cap 5-15 cm. Aug.-Nov.

121

Cap fungi with gills

Milk Caps, pages 122-3, have crumbly flesh and white or pale yellow gills and spores. They ooze white or coloured milky drops when broken.

Cream-pink gills run down stem

White milky drops

Woolly cap

Cap is funnel-shaped when mature

Woolly Milk Cap ▶

Lactarius torminosus
White flesh oozes white juice when cut. Grows in mixed woods and on heaths, especially near birch. Poisonous.
Cap 5-12 cm.
Sept.-Nov.

Hollow stem

Green milky drops

Saffron Milk Cap ▶

Lactarius deliciosus
Orange to fawn, funnel-shaped cap with light and darker rings. Oozes green milk when cut. Grows in coniferous woods.
Cap 4-10 cm.
Aug.-Nov.

Turns green when bruised

Cap fungi with gills

Peak at centre of cap

Gills run down stem

White milky drops

Lactarius rufus
Cap at first covered with downy hairs, later smooth. Cream coloured gills when young, turning pale red-brown with age. Under pines. Cap 3-10 cm. Aug.-Nov.

White Milk Cap ►
Lactarius vellereus
Cap depressed at centre and feels velvety. Yellowy-white gills run down stem and may be tinged brown. Often in groups in broad-leaved woods. Cap 8-20 cm. Sept.-Nov.

Edge of cap turns under

White milk

Short thick stem

Short thick stem is sticky when wet

Brown milky drops

◄ **Ugly Milk Cap**
Lactarius turpis
Cap covered with thick yellow down when young. White flesh turns violet-grey when cut. Cream gills have brown edges. Under conifers and birch. Cap 8-24 cm. Aug.-Nov.

Cap fungi with gills

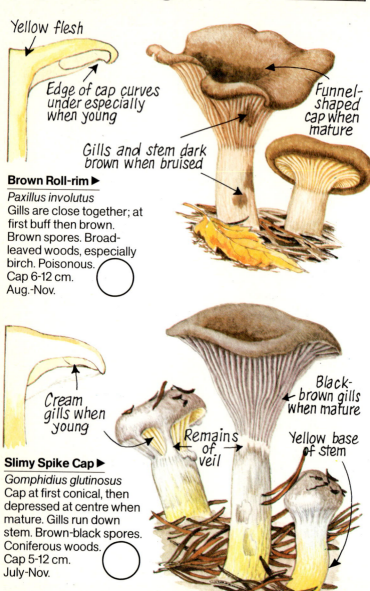

Yellow flesh

Edge of cap curves under especially when young

Gills and stem dark brown when bruised

Funnel-shaped cap when mature

Brown Roll-rim ▶

Paxillus involutus
Gills are close together; at first buff then brown. Brown spores. Broadleaved woods, especially birch. Poisonous. Cap 6-12 cm. Aug.-Nov.

Cream gills when young

Black-brown gills when mature

Remains of veil

Yellow base of stem

Slimy Spike Cap ▶

Gomphidius glutinosus
Cap at first conical, then depressed at centre when mature. Gills run down stem. Brown-black spores. Coniferous woods. Cap 5-12 cm. July-Nov.

Funnel-shaped fungi

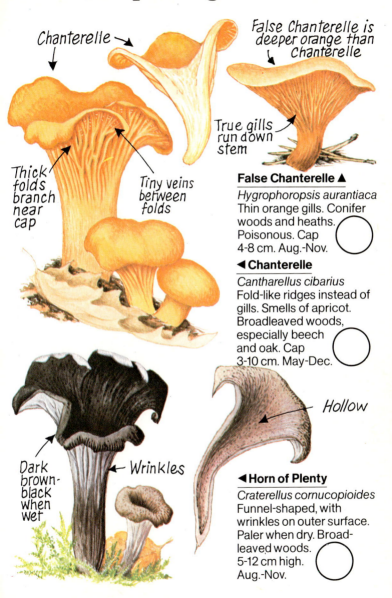

Chanterelle

False Chanterelle is deeper orange than Chanterelle

True gills run down stem

Thick folds branch near cap

Tiny veins between folds

False Chanterelle ▲
Hygrophoropsis aurantiaca
Thin orange gills. Conifer woods and heaths. Poisonous. Cap 4-8 cm. Aug.-Nov.

◄ Chanterelle
Cantharellus cibarius
Fold-like ridges instead of gills. Smells of apricot. Broadleaved woods, especially beech and oak. Cap 3-10 cm. May-Dec.

Hollow

Dark brown-black when wet

Wrinkles

◄ Horn of Plenty
Craterellus cornucopioides
Funnel-shaped, with wrinkles on outer surface. Paler when dry. Broad-leaved woods. 5-12 cm high. Aug.-Nov.

Cap fungi with spines

The fungi on this page have a mass of spines, instead of gills or pores, underneath the cap. The spores are produced in these spines.

Creamy pink spines

Wood Hedgehog ▶

Hydnum repandum
Cream to light brown cap.
Thick stem is narrower at
base. Often grows in rings
in broadleaved woods,
especially under beech
and oak.
Cap 5-15 cm.
Aug.-Nov.

Grey flesh

Dark grey
spines run
down stem

Sarcodon imbricatum ▶

Grey-brown cap with
darker overlapping scales.
Coniferous woods, usually
on poor sandy soil.
Cap 5-20 cm.
Sept.-Nov.

Bracket fungi with gills

Bracket fungi with gills grow on trees (living wood) or stumps (dead wood). The gills run down a short stem that is usually to one side of the cap.

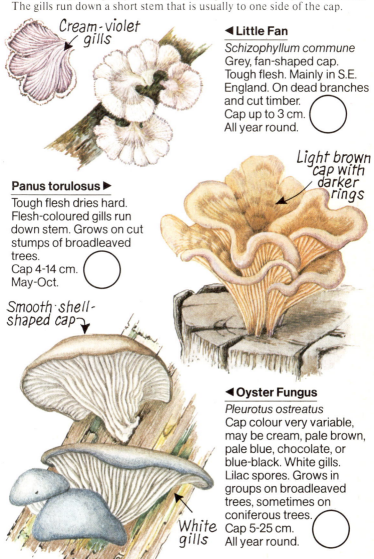

Cream-violet gills

◀ Little Fan
Schizophyllum commune
Grey, fan-shaped cap.
Tough flesh. Mainly in S.E.
England. On dead branches
and cut timber.
Cap up to 3 cm.
All year round.

Light brown cap with darker rings

Panus torulosus ▶
Tough flesh dries hard.
Flesh-coloured gills run
down stem. Grows on cut
stumps of broadleaved
trees.
Cap 4-14 cm.
May-Oct.

Smooth shell-shaped cap

◀ Oyster Fungus
Pleurotus ostreatus
Cap colour very variable,
may be cream, pale brown,
pale blue, chocolate, or
blue-black. White gills.
Lilac spores. Grows in
groups on broadleaved
trees, sometimes on
coniferous trees.
Cap 5-25 cm.
All year round.

White gills

Bracket fungi with pores

The Bracket fungi with pores, pages 128-30, all grow on trees or stumps and have masses of fine tubes, opening as small pores on the undersurface of the cap. The spores are produced in these tubes.

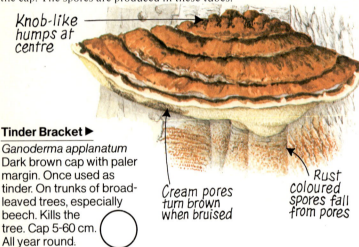

Knob-like humps at centre

Tinder Bracket ▶

Ganoderma applanatum
Dark brown cap with paler margin. Once used as tinder. On trunks of broad-leaved trees, especially beech. Kills the tree. Cap 5-60 cm. All year round.

Cream pores turn brown when bruised

Rust coloured spores fall from pores

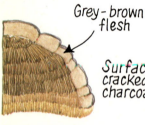

Grey-brown flesh

Surface is cracked like charcoal

Tinder Fungus ▶

Phellinus igniarius
Very hard, grey to black-brown, cracked cap. Tiny pale grey pores turn light brown with age. On broad-leaved trees, especially willow and poplar. Cap 10-30 cm. All year round.

Pale margin

Bracket fungi with pores

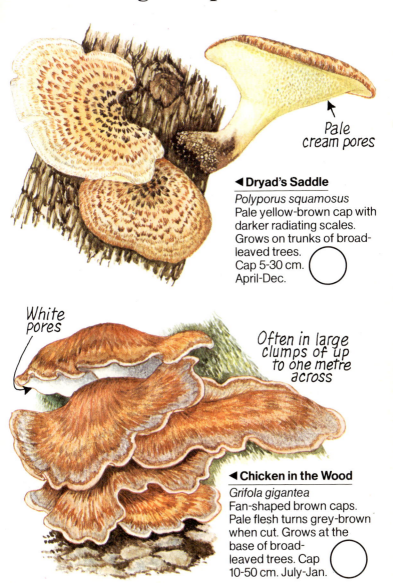

Pale
cream pores

◄ Dryad's Saddle

Polyporus squamosus
Pale yellow-brown cap with
darker radiating scales.
Grows on trunks of broad-
leaved trees.
Cap 5-30 cm.
April-Dec.

White
pores

Often in large
clumps of up
to one metre
across

◄ Chicken in the Wood

Grifola gigantea
Fan-shaped brown caps.
Pale flesh turns grey-brown
when cut. Grows at the
base of broad-
leaved trees. Cap
10-50 cm. July-Jan.

Bracket fungi with pores

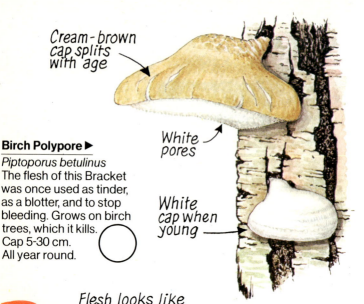

Cream-brown cap splits with age

White pores

White cap when young

Birch Polypore ▶

Piptoporus betulinus
The flesh of this Bracket was once used as tinder, as a blotter, and to stop bleeding. Grows on birch trees, which it kills.
Cap 5-30 cm.
All year round.

Flesh looks like raw meat

Pale red-brown pores

Beef Steak Fungus ▶

Fistulina hepatica
Soft, fleshy fungus that oozes red drops when squeezed. Pores at first yellow then pale red-brown with age. At base of broad-leaved trees, especially oak and chestnut.
Cap 5-30 cm.
Aug.-Nov.

Small brackets and crust fungi

Many fungi form crusts on twigs, trunks, logs or on the ground. Some of these, such as *Trametes serialis,* have pores. Others, like *Stereum hirsutum,* produce spores over their flat or wrinkled undersurface.

Rings of colour on cap

Paler at edge

◄ Coriolus versicolor

Brightly-coloured bracket with velvet-like cap. Grows in layers on cut stumps and branches of broadleaved trees. Cap 2-5 cm wide. All year round.

Small pores under cap

Trametes serialis ►

Cream-coloured with small pores all over surface. Forms a crust on conifer tree trunks and stumps. All year round.

Close-up showing tiny hairs

◄ Stereum hirsutum

Smooth yellowish upper surface, covered with tiny hairs. Very common on stumps, branches and cut wood. Causes white rot on stored timber. 1-4 cm across. All year round.

Stinkhorn, Saddle Cap

Cross-section of "egg"

Spores

"Egg"

Stinkhorn ▶

Phallus impudicus
Develops from a white "egg", buried in leaves or moss. Dark slime on head of stem contains spores and is soon eaten by flies. Very strong, unpleasant smell. Grows in woods and hedgerows. 6-12 cm high. July-Oct.

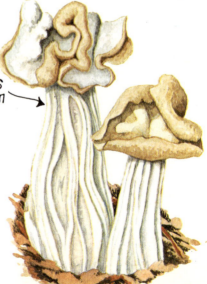

Deep grooves on stem

Saddle Cap ▶

Helvella crispa
Hollow stem is capped by folds that are off-white on the upper side and pale fawn underneath. Beside paths in broadleaved woods. 5-10 cm high. March, April and Aug.-Oct.

Morel, Turban Fungus

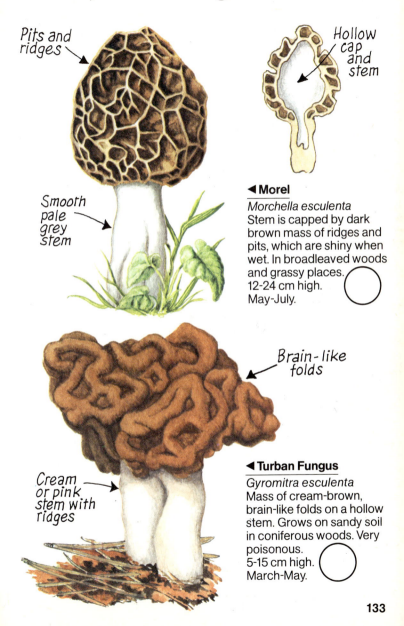

Pits and ridges

Hollow cap and stem

Smooth pale grey stem

◄ Morel
Morchella esculenta
Stem is capped by dark brown mass of ridges and pits, which are shiny when wet. In broadleaved woods and grassy places. 12-24 cm high. May-July.

Brain-like folds

Cream or pink stem with ridges

◄ Turban Fungus
Gyromitra esculenta
Mass of cream-brown, brain-like folds on a hollow stem. Grows on sandy soil in coniferous woods. Very poisonous. 5-15 cm high. March-May.

Cup fungi

Cup fungi grow on logs, tree trunks or on the ground. Their spores are produced from the inside of the cup.

Scarlet Elf Cup ▶

Sarcoscypha coccinea
Smooth, scarlet surface inside cup; downy, cream or orange outer surface. On rotting branches of broadleaved trees. Cup 2-5 cm wide. Dec.-March.

Cup flattens with age

◀ Peziza badia

Outside surface of cup is paler brown than inside. On the ground, in broad-leaved woods. Poisonous. Cup 3-7 cm wide. Aug.-Oct.

Edge of cup often splits with age

Orange Peel Fungus ▶

Aleuria aurantia
Smooth, orange surface inside cup; downy, paler orange outer surface. On gravel, lawns and on bare soil in woods. Cup 1-12 cm wide. Aug.-Dec.

Cup fungus, Bird's Nest Fungus, Truffle

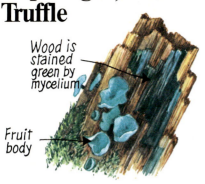

Wood is stained green by mycelium

Fruit body

◀ **Blue Stain Fungus**

Chlorosplenium aeruginascens
Tiny, green cup-shaped fruit bodies. Wood on which it grows is permanently stained green. Grows on rotting wood, especially oak. Cup 0.5 cm wide. May-Nov.

Bird's Nest Fungus ▶

Crucibulum vulgare
Small cup-shaped fungus, filled with several egg-shaped bodies which contain spores. Rain splashes the "eggs" out of the cup and the spores can then disperse. On twigs.
Cup 0.5-1 cm high.
Sept.-Feb.

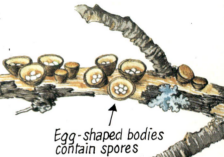

Egg-shaped bodies contain spores

Marbled flesh

Rough warty surface

◀ Cook's Truffle

Tuber aestivum
This is one of several kinds of truffles. Grey flesh, sometimes tinged lilac, streaked with paler veins. Grows in soil, just below ground level, in beech woods, especially in chalk soil.
3-8 cm across.
Aug.-Oct.

Puffballs

Puffballs are round or pear-shaped fungi with a skin that is either smooth or covered with small warts. Their spores are produced inside the "ball" and, when ripe, are dispersed either through a small hole that forms in the top of the ball, or through cracks over the surface.

Flesh at first white then yellow

Spores released through hole

Common Puffball ▶
Lycoperdon perlatum
Covered with small warts. White flesh turns yellow-green as spores ripen. Small hole forms at top, when mature. In woods. 4-7 cm high. July-Nov.

Giant Puffball ▶
Calvatia giganteum
One of the world's largest fungi. Flesh at first white then yellow. Splits when mature. In fields, woods and hedgerows. 15-100 cm across. Aug.-Nov.

136

Puffball, Earthball, Earth Star

◀ Puffball

Calvatia caelatum
Grey-brown at first, turning
dark brown with age. Top
opens out to release dark
brown spores. In fields and
woods on sandy
soil. 6-10 cm
across. July-Nov.

Black
centre
when
mature

Common Earthball ▶

Scleroderma aurantium
Looks like a young
puffball, but is much
harder and no hole forms.
Centre is at first cream,
then yellow and finally
black. In woods, especially
under birch.
4-8 cm across.
July-Dec.

← Hard
warty
surface

Onion-shaped
at first

Note:
There are
other kinds of
earthballs
and earthstars

"Collar"
round
centre

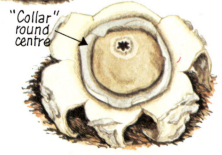

◀ Earth Star

Geastrum triplex
Onion-shaped at first, then
outer layer peels back
forming 5-7 pointed arms.
Broadleaved woods,
especially beech.
6-8 cm across.
Aug.-Nov.

Jelly fungi

Many fungi have a jelly-like texture, especially when wet. Some jelly fungi are irregular in shape and grow in clusters.

Witch's Butter ▶

Exidia glandulosa
Deep olive to black coloured jelly. Grows in mass of varying size on rotting stumps of broad-leaved trees. All year round, but mainly Oct.-Dec.

Brain-like folds

◀ Yellow Brain Fungus

Tremella mesenterica
Slimy when wet, drying to a hard, orange crust. Grows in mass of varying size on dead branches and tree stumps. All year round, but mainly Sept.-Dec.

Dries to a hard crust

Soft and velvety when wet

Ear Fungus ▶

Auricularia auricula
Mainly on branches of elder trees. Each ear 3-10 cm across. All year round.

Paler brown outer surface

Cauliflower Fungus, Fairy Clubs

Fairy Clubs may be branched, like the Coral Fungus, or unbranched, like the Giant Club. They have a leathery texture and, like some of the jelly fungi, produce their spores all over their surface.

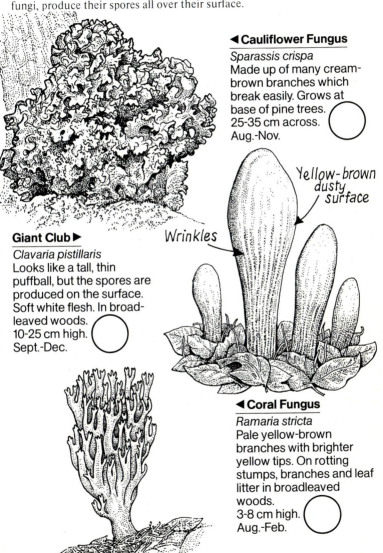

◀ Cauliflower Fungus
Sparassis crispa
Made up of many cream-brown branches which break easily. Grows at base of pine trees.
25-35 cm across.
Aug.-Nov.

Yellow-brown dusty surface

Wrinkles

Giant Club ▶
Clavaria pistillaris
Looks like a tall, thin puffball, but the spores are produced on the surface. Soft white flesh. In broad-leaved woods.
10-25 cm high.
Sept.-Dec.

◀ Coral Fungus
Ramaria stricta
Pale yellow-brown branches with brighter yellow tips. On rotting stumps, branches and leaf litter in broadleaved woods.
3-8 cm high.
Aug.-Feb.

139

CHAPTER 4
Lichens

Two in one
Lichens are small plants, but they are also some of the most unusual members of the plant kingdom. It is only just over a hundred years since scientists discovered that a lichen is not a single plant, but two different plants living together in a relationship called symbiosis. A symbiotic relationship is one that benefits both partners. The two plants in a lichen symbiosis are a fungus and an alga. You will find pictures of large fungi in Chapter 3 and of seaweeds, which are large algae, in Chapter 5.

Structure
The body of a lichen is not divided into separate parts which all do different jobs, as do the leaves, stems and roots of trees or flowers. The body of a lichen is called a thallus, and in most lichens, the greater part of the thallus is made up of fine threads of fungus called hyphae. This provides a rigid framework within which single cells of alga are embedded. In many lichens some of the fungus hyphae form fine threads by which the plant is attached to the surface. These are called rhizines.

Types of lichen
There are four main forms of lichen shown in this book:

1. Branch-like lichens (p.142)
These are branched and bushy, and may be either upright or hanging. They are usually attached to the surface at only one point.

2. Leaf-like lichens (p.143)
These usually grow horizontally and have a thallus which looks like leaves or scales. It is attached to the surface by rhizines.

3. Cup lichens (p.144)
Cup lichens are very distinctive. They have erect cup-like structures which arise from a surface mat of scales.

4. Crust lichens (p.145)
These grow flat against the surface on which they live. This can be rock, wood or soil. They are firmly attached and may even be embedded within the surface.

Finding and identifying lichens

Colour is one obvious way of identifying lichens. They can be yellow, orange, red, grey, brown, black or greyish-green, and some patches of lichen may contain more than one colour. Remember that the colour will depend on whether the lichen is wet or dry.

Look for lichens on walls, rocks, roofs, gravestones, paths, fence posts, tree trunks, amongst turf on heathland and on the soil of wayside banks. Some species occur only on a particular surface such as rock or wood and this may help you to name them.

There are very few lichens that survive in the polluted air of cities – although recently lichens have been returning, probably because city air is cleaner than it used to be.

How lichens live

The algal part of the lichen contains chlorophyll and may be blue-green or grass green. The chlorophyll enables the alga to use the sun's energy to make its own food from carbon dioxide and water. It can supply not only its own needs but also those of the surrounding fungus which has no chlorophyll and cannot manufacture food in this way. In return, the fungus provides shelter for the alga. It also extracts water and mineral salts from the surface on which it grows.

How lichens reproduce

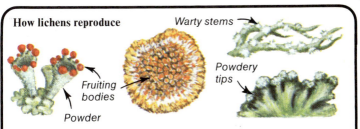

Warty stems

Fruiting bodies

Powder

Powdery tips

▲ Some examples of reproductive structures you might see.

Although lichens grow slowly and live for a long time, reproduction and distribution is probably just as important as in other plants. Lichens have several different methods of doing both.

On the surface of some lichens you may see a powdery substance. This consists of tiny granules each of which contains enough fungus and alga to grow and establish a new plant. They are very small and light and can be blown a long way.

The fungus part of the lichen sometimes produces spores in special fruiting bodies. The success of this method depends on the fungal spore meeting the correct alga with which it can combine to produce a copy of the parent plant. Many of the growths, bumps and different coloured areas on the surface of lichens are to do with reproduction.

Sometimes a piece of dry lichen breaks off and is blown away to a new place. If conditions there are right, a new plant may grow.

Branch-like lichens

Ragged Mealy Lichen ▶

Usually grows on trees.
Fruiting bodies are rare.
The edges of the branches
are covered with tiny
powdery plates.
Tufts about
3 cm across.

Powdery
plates

Flat branches

Branches are rough,
warty and
cylindrical

Black
base

◀ Beard Lichen

A branched lichen which
forms tangled tufts. Its
branches start erect but
soon hang down. Usually
found on trees.
Tufts about
7 cm across.

Upper surface is
wrinkled and
covered in
powder

Branches are
greenish-grey
on one side,
white on the other

Oak Moss ▶

The flattened and forked
branches of this lichen
form shaggy tufts. The
branches are criss-crossed
by ridges. Found
on trees and
fences. Tufts
4-5 cm across.

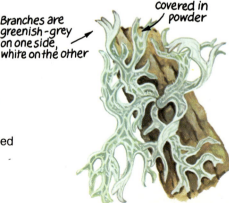

Leaf-like lichens

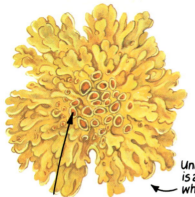

Fruiting body

◄ Common Orange Lichen

An easily-spotted lichen which forms patches on roofs, walls, tombstones and rocks, especially by the sea. The edge is slightly raised. Patches about 15 cm across.

Underside is almost white

Fruiting body

Downy upperside

Lobe

Dog Lichen ►

A large lichen with lobes that curl up at the edges. When moist, it is dark brown on top; when dry, a pale grey. Grows on turf, earthy walls and decaying logs. Lobes 3 cm wide.

White hairlike threads on underside

Lobes are hollow and white inside, when cut

◄ Puffed Lichen

A smooth, grey-green lichen. The narrow lobes have an inflated appearance. Grows on trees and rock surfaces, sometimes near towns. Branches 1-2 cm long.

White powder on swollen tips

Tips turn up

143

Cup lichens

Cock's Comb Lichen ▶

Forms patches of scales from which arise cup-like structures. On the rims of these are tiny red fruiting bodies. Heath and moorland. Up to 5 cm high.

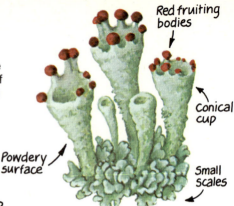

Red fruiting bodies

Conical cup

Powdery surface

Small scales

Wide top to cup

Greenish powder on surface

Mat of small scales

◀ Slender Pixie-cup Lichen

Forms mats of tiny scales which are white underneath. The cups are borne on slender stems. Grows on sand dunes, tree roots, walls and banks. Up to 5 cm high.

Devil's Matches ▶

Erect stalks arise out of patches of leafy scales. These carry bright red fruiting bodies, and look like matchsticks. Common on heaths and moors. Up to 2 cm high.

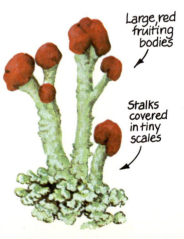

Large, red fruiting bodies

Stalks covered in tiny scales

Crust lichens

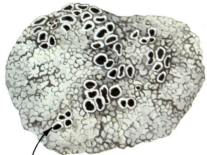

Fruiting bodies are purple inside when cut

◀ Black Shield Lichen

Forms thick, warty, crust-like patches on rocks, walls and tombstones, especially by the sea. The fruiting bodies have black centres. Patches 5-10 cm across.

Grey Crust Lichen ▶

The commonest crust lichen of town centres. It forms an inconspicuous thick crust with no definite edges. Found on walls, pavements, tombstones.

Thick grey-green crust has a powdery surface

Fruiting bodies are dark yellow

◀ Caloplaca heppiana

Forms discs, often inter-locking. Differs from Common Orange Lichen in being crust-like rather than leafy and is a less intense shade of orange. Found on walls and tombstones. Up to 5 cm across.

CHAPTER 5

Seaweeds

The main parts of a seaweed
Seaweeds belong to a large group of very simple plants called algae. They vary greatly in form but most consist of a frond (a leaf-like structure), a stalk (called a stipe), and a holdfast (by which the plant is attached to the surface upon which it grows). The frond, stipe and holdfast are together called the thallus.

How to identify a seaweed
The first point to notice about a seaweed is its colour. There are three types of seaweed: the browns, the greens and the reds. Most brown seaweeds are brown or olive green in colour. They are the most obvious of all the seaweeds around our coasts. Some of them, such as Kelp, are very large. The green seaweeds are less common than the browns and are usually a bright grass green. The largest group of seaweeds, the reds, are common on rockier shores.

The second point to notice is the shape of the thallus. Is it branched or unbranched? Are the branches rounded or flattened? Do they have a central rib? Some seaweeds have air bladders which help keep the thallus upright in the water. Try putting the seaweed you have found back in the water in order to see its shape properly.

Where to look
Seaweeds, like other plants, need light in order to live and cannot survive in waters so deep that light cannot get to them. You are most likely to find green algae growing

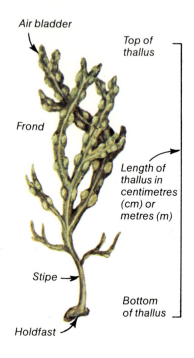

Air bladder

Top of thallus

Frond

Length of thallus in centimetres (cm) or metres (m)

Stipe

Bottom of thallus

Holdfast

high up on rocky shores. Look for the brown seaweeds growing lower down on the shore. Most red seaweeds grow at greater depths but you will often find all types of seaweed washed up on the beach after a high tide or a storm.

The description next to each illustration tells you where each species normally grows. If you find a seaweed still attached to a surface, note which part of the beach it is growing on. This will help you to identify it.

Holdfasts

Holdfasts vary considerably in shape and can be a useful aid to identification. Although some look very like the roots of other plants, they are not actually roots. Holdfasts are solely a means of anchoring the seaweed. In this they differ from the roots of other plants which also take up water and mineral salts for the plant. Holdfasts may vary depending on the kind of surface to which the seaweed is attached.

How do seaweeds reproduce?

In some seaweeds, special reproductive structures appear at certain times of year. Their location and form can sometimes be a good way of identifying the plant. In the brown wracks, these reproductive structures take the form of obvious swellings at the tips of the branches. These are solid, and should not be confused with air bladders, which are hollow. In many other seaweeds, the reproductive structures are small and less obvious.

The means by which seaweeds reproduce and disperse are often complex, involving a male and a female plant and two separate stages in the life cycle. Sometimes, however, a new plant can arise from a piece of the parent plant which has broken away.

The zones of the seashore

Twice every day the sea comes up the beach and goes down again.

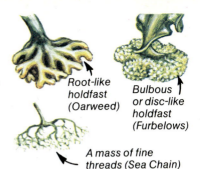

Root-like holdfast (Oarweed)

Bulbous or disc-like holdfast (Furbelows)

A mass of fine threads (Sea Chain)

▲ Three types of holdfast.

These movements of the sea are known as tides and are caused by the attraction of the sun and the moon. Every day there are two high tides and two low tides. About every two weeks there are very high and very low tides called **spring tides.** Between the spring tides, also at fortnightly intervals, there are small tides called **neap tides.** Between spring and neap tides, the tides steadily increase and decrease. The local paper will tell you the state of the tides.

The area between the low water lines of the spring and neap tides is called the **lower shore.** The area between the high water lines of the spring and neap tides is called the **upper shore.** The main part of the beach between these two is called the **middle shore.** The descriptions of seaweeds refer to these three areas.

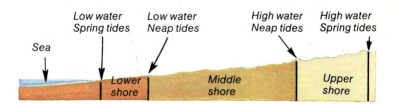

Low water Spring tides

Low water Neap tides

High water Neap tides

High water Spring tides

Sea

Lower shore

Middle shore

Upper shore

Gut Laver (left) ▶

Tube-like fronds do not branch. May cover pools on upper shore and in estuaries. Very common.
20 cm.

Sea Lettuce (right) ▶

Common on rocky shores at middle and lower levels. Fronds become dark green with age.
20 cm.

Frond

Disc

◀ Mermaid's Cup

Disc shape on thin stalk made up of many tiny segments pressed close together. On rocks in sheltered bays. Mediterranean.
4-6 cm tall.

Bryopsis (left) ▶

Looks shiny. Found on steep sides of rock pools on middle and lower shore.
7.5 cm.

Sea Chain (right) ▶

Feels hard and brittle because it is covered with lime. Shallow water in sheltered bays. Mediterranean.
15 cm.

Female plant is darker green

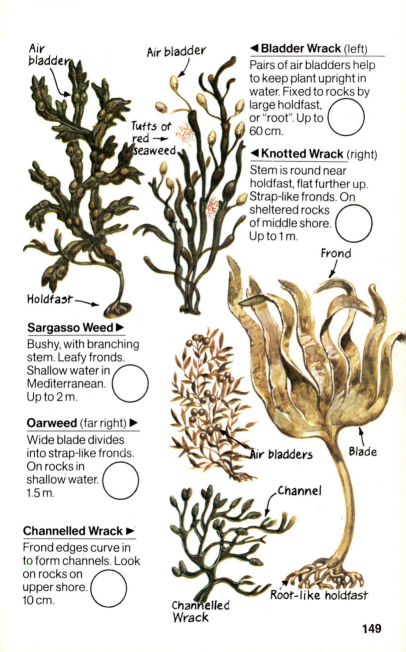

Air bladder

Air bladder

Tufts of red → seaweed

Holdfast →

◄ Bladder Wrack (left)
Pairs of air bladders help
to keep plant upright in
water. Fixed to rocks by
large holdfast,
or "root". Up to
60 cm.

◄ Knotted Wrack (right)
Stem is round near
holdfast, flat further up.
Strap-like fronds. On
sheltered rocks
of middle shore.
Up to 1 m.

Frond

Sargasso Weed ►
Bushy, with branching
stem. Leafy fronds.
Shallow water in
Mediterranean.
Up to 2 m.

Oarweed (far right) ►
Wide blade divides
into strap-like fronds.
On rocks in
shallow water.
1.5 m.

Channelled Wrack ►
Frond edges curve in
to form channels. Look
on rocks on
upper shore.
10 cm.

Air bladders

Blade

Channel

Channelled
Wrack

Root-like holdfast

149

Phymatolithon ▶

Some red seaweeds, like this one, have a hard coating of lime. It forms a crust in patches on rocks and stones on middle and lower shore.

◀ Laver

Bumpy fronds usually attached at one point. On sand-covered stones, middle to lower shore. Rocks on upper shore. 15 cm.

Plocamium ▶

Small tufted plant with finely-divided fronds. Feathery tips only grow on one side of each branch. Shallow water or washed ashore. 15-20 cm.

◀ Irish Moss

Two forms, broad and narrow, found on rocks on middle and lower shore. Look for the small, disc-shaped holdfast. 15 cm.

Narrow form

Broad form

Holdfast

Sea Oak ▶

Fronds shaped like oak leaves, with markings like veins. Grows on lower shore rocks, in pools and on stalks of large brown seaweeds. 20 cm.

Stalk of brown seaweed

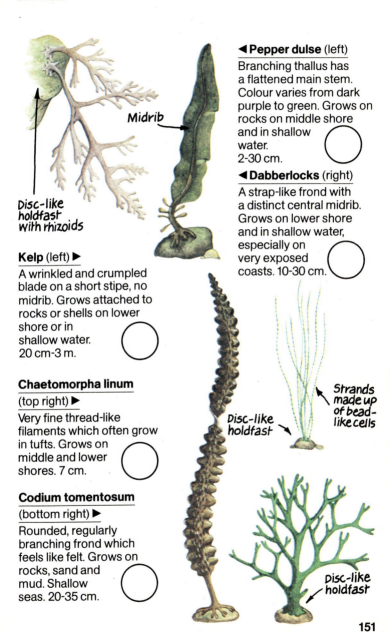

Midrib

Disc-like holdfast with rhizoids

Kelp (left) ▶

A wrinkled and crumpled blade on a short stipe, no midrib. Grows attached to rocks or shells on lower shore or in shallow water. 20 cm-3 m.

Chaetomorpha linum

(top right) ▶

Very fine thread-like filaments which often grow in tufts. Grows on middle and lower shores. 7 cm.

Codium tomentosum

(bottom right) ▶

Rounded, regularly branching frond which feels like felt. Grows on rocks, sand and mud. Shallow seas. 20-35 cm.

◀ Pepper dulse (left)

Branching thallus has a flattened main stem. Colour varies from dark purple to green. Grows on rocks on middle shore and in shallow water. 2-30 cm.

◀ Dabberlocks (right)

A strap-like frond with a distinct central midrib. Grows on lower shore and in shallow water, especially on very exposed coasts. 10-30 cm.

Disc-like holdfast

Strands made up of bead-like cells

Disc-like holdfast

CHAPTER 6

Mosses and Liverworts

Mosses

There are over 600 different kinds of mosses growing in Britain. They all have clearly defined stems, and leaves which have no stalks. They are attached to the surface upon which they grow by fine root-like threads called rhizoids. The leafy shoots may be long and creeping, or short and erect. They vary in length from less than a centimetre to over 20 centimetres

How to identify mosses

Mosses grow in many different forms – some look like deep shaggy carpets, others form neat little cushions, and yet others are smooth and shiny like velvet.

All mosses are green, but the shade varies from light bright green to pale whitish blue-green. Remember that the colour often varies with the type of soil and may also depend on whether the moss is wet or dry.

Look at the way the leaves are arranged on the stem. Are they in two rows or do they form a spiral around the stem? The leaves of some mosses have a thickened midrib or ridge; others have a fine hair-like point.

The moss you find may be fruiting; if so, look carefully at the capsule and compare it with the illustrations in this Chapter.

Where to look

You will find mosses on rocks, walls, roofs, trees or on the ground. You will find them in cities, high up on mountains, in bogs, by streams and carpeting woodland floors.

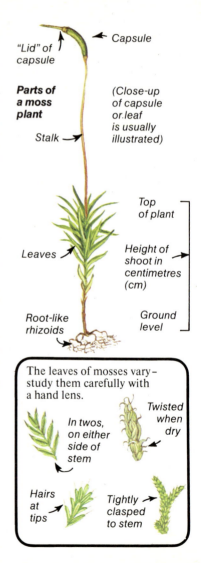

Capsule

"Lid" of capsule

Parts of a moss plant

(Close-up of capsule or leaf is usually illustrated)

Stalk

Leaves

Root-like rhizoids

Top of plant

Height of shoot in centimetres (cm)

Ground level

The leaves of mosses vary – study them carefully with a hand lens.

In twos, on either side of stem

Twisted when dry

Hairs at tips

Tightly clasped to stem

The life cycle of a moss

The male and female sex organs of mosses may be located in different plants or in the same plant (either on the same shoot or on different shoots). Sperm cells released by the male sex organ swim in dew or rainwater to the female sex organ, which contains the egg. The egg, once fertilized, grows into a new plant which is still attached to the female parent. Only the capsule and stalk of the new plant can be seen, on top of the parent plant.

The capsule contains very small spores, which are released, and then dispersed by the wind. If they come to rest in a suitable place, they will germinate to form fine threads of green cells. From this, new leafy shoots will develop, bringing the moss back to the start of the life cycle.

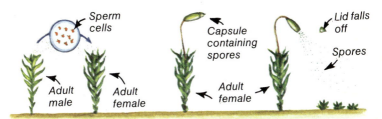

Sperm cells released from male plant fertilize eggs in female plant.

Capsule grows from fertilized female plant.

Spores are released from capsule and grow into new young plants.

Liverworts

There are two main kinds of liverworts. The thallose liverworts (see illustration on right) have no distinct stems or leaves. The branched body is thin and leathery and grows flat upon a surface. This body is called a thallus.

Leafy liverworts have both stems and leaves – hence their name. The leaves are usually in two rows along the stem and there is sometimes a third row of under-leaves. The leaves of these liverworts always have two lobes, although you may have to look carefully and use a hand lens in order to see them. One is usually smaller than the other and the smaller lobe is often turned back upon the larger lobe.

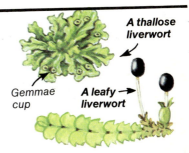

Another method of reproducing

Some mosses and many liverworts increase their number by producing special structures called gemmae. These gemmae are formed on the moss or liverwort, often in little cups, called gemmae cups. The gemmae are dispersed by raindrops and grow directly into a new plants.

Bog Moss ▶

Forms large bright green or whitish cushions on bogs. Dead leaves and stems accumulate to form peat. Grows on acid soils. Up to 25 cm.

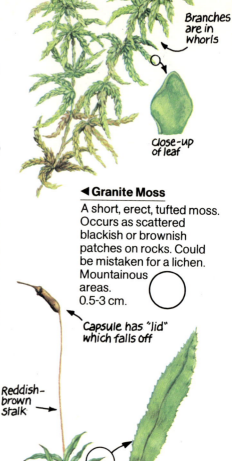

Branches are in whorls

close-up of leaf

Capsule is almost hidden

Close-up of capsule – notice its four openings

Small, overlapping, close-set leaves

◀ Granite Moss

A short, erect, tufted moss. Occurs as scattered blackish or brownish patches on rocks. Could be mistaken for a lichen. Mountainous areas. 0.5-3 cm.

Capsule has "lid" which falls off

Reddish-brown stalk

Leaf is wavy when moist

Catherine's Moss ▶

An obvious and very attractive plant with upright branching stems. Leaves become twisted when dry. Common in woods and on heaths. 5-7 cm.

Narrow, pale border at edge of leaf

Leaves in twos

◄ Fissidens bryoides

A small but fairly common moss which forms dense colonies on banks and disturbed ground. Shoots look like tiny fern fronds. 1-1.5 cm.

Short "beaked" lid

Box-like capsule

When young, capsules are covered in silky brown hairs

Hair Moss ►

A tall luxurious plant with tough upright stems. Forms large, bright green tufts and is abundant on moors and bogs, and by streams. 15-40 cm.

Toothed edge

Notice teeth at end of capsule

Lid has fallen off

Mature capsule is almost horizontal. When young, it is more upright.

◄ Ceratodon purpureus

A common moss which is particularly obvious in spring with its reddish capsules and their purple stems. Grows on bare ground. 2-3 cm.

Leucobryum glaucum ▶

Remarkable for its very pale colour. Cushions are only loosely attached to ground and sometimes break away, forming "moss balls". Found in woods and on bogs. 10-15cm.

Close-up of one stalk

Grows in dense, humped cushions

Teeth can untwist to allow release of spores from capsule

Close-up of capsule

Leaves have hairs at their tips

◀ Tortula muralis

Forms small cushions which appear grey when dry. When moist, leaves form neat rosettes; when dry, they become curled, and look similar to those of *Barbula convoluta*, but with hairs. Common on walls. Up to 1 cm.

Pale yellow stem

Leaves curl up when dry; unlike those of Tortula muralis (above), leaves have no hairs

Barbula convoluta ▶

A short-stemmed moss which forms small mats or cushions on top of walls and on bare ground. Leaves are small and blunt. 1 cm.

◄ Fire Moss

Stems are short and unbranched. Capsule stalks become curved with age, making plant look tangled. Often grows on ground which has recently been burnt. 1-1.5 cm.

Mouth of capsule has spiralled teeth which control the release of spores

Leaves have whitish, hair-like points

Toothed edge

Rhacomitrium
lanuginosum ►

An obvious and much-branched moss, common on mountain tops, where it forms extensive greyish mats known as "Rhacomitrium heath". 12-25 cm.

Capsule

Leaves curl up in spirals when dry

◄ Bryum capillare

A very common moss which forms green or reddish patches on rocks, trees and the tops of walls. Capsule turns brown when ripe and has an orange lid. 2-5 cm.

Leaves have long, fine points

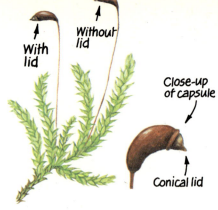

Brachythecium rutabulum ▶

Grows in large, straggly tufts. Glossy green or yellow branches arise irregularly from creeping stems. Common in woods or shady places. Up to 5 cm.

With lid

Without lid

Close-up of capsule

Conical lid

Close-up of tip of shoot

Blunt tips

◀ Pseudoscleropodium purum

A large plant which forms loose tufts. Is abundant on chalk grasslands, but also occurs in other open situations. Leaves overlap each other. 15 cm.

Capsule

"Beaked" lid

Curved leaves overlap tightly

Hypnum cupressiforme ▶

Long, regularly branched stems grow along the ground forming dense carpets. Common on stones, walls, trees and soil in gardens and woods. Capsules often produced in winter and spring. 1 cm.

Thallus of shiny, dark green lobes

Black capsule

◄ Pellia epiphylla
A thallose liverwort. Grows in moist places by ditches and streams, also on peat. Thallus about 1 cm across.

Gemmae cup

Marchantia polymorpha ►
A large thallose liverwort. Produces male and female reproductive structures in spring. Common on wet moorland, also in green-houses. Thallus up to 10 x 20 cm.

Open, ripe capsule

9-rayed female structure

Disc-like male structure

Unripe capsule

Sheath

◄ Lophocolea heterophylla
A common leafy liverwort with two rows of leaves and a third row of less obvious underleaves. Grows on decaying logs and tree stumps in damp woodland. Branches 1-2 cm long.

CHAPTER 7

Ferns

The parts of a fern
The part of the fern which you can see above ground is the leaf, called the frond. The stem is underground and is called the rhizome. The roots grow from the rhizome.

Although the fronds are usually killed by the first frost, the rhizome and roots remain alive through the winter. Young fronds, called fiddle-heads or croziers, arise from the rhizome and the roots the following year.

Where to look
The greatest variety of ferns are found in the tropical areas of the world, where some grow into tree forms more than twelve metres high. In this country, ferns are much smaller. You will find them in damp and shady places, in woods, on rocks and on walls.

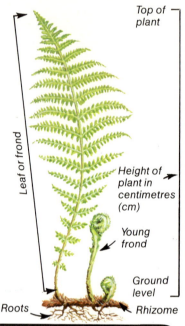

Top of plant

Leaf or frond

Height of plant in centimetres (cm)

Young frond

Ground level

Roots

Rhizome

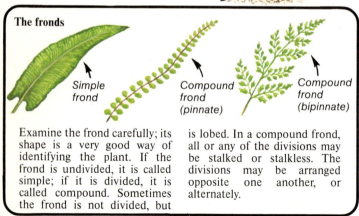

The fronds

Simple frond

Compound frond (pinnate)

Compound frond (bipinnate)

Examine the frond carefully; its shape is a very good way of identifying the plant. If the frond is undivided, it is called simple; if it is divided, it is called compound. Sometimes the frond is not divided, but is lobed. In a compound frond, all or any of the divisions may be stalked or stalkless. The divisions may be arranged opposite one another, or alternately.

How ferns spread

Ferns are dispersed to new places by means of tiny wind-blown spores. These are produced in sori – the brownish dots which appear on the underside of the fronds in late spring or early summer.

Ferns have a complex life cycle. Spores are released from the sori and, if they come to rest in suitable conditions, germinate to form a heart-shaped structure called a prothallus. The prothallus is very small, about 6 cm long, but it is visible to the naked eye.

The male and female reproductive structures are on the underside of the prothallus. When moisture is plentiful, the sperm swim from the male structures to the female structures – usually on a different prothallus. The egg in the female structure is fertilized and a new fern plant grows up from the prothallus.

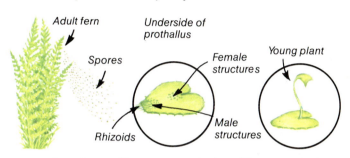

Adult fern
Spores
Underside of prothallus
Female structures
Young plant
Rhizoids
Male structures

Spores are released from sori on the undersides of the fronds.

The spores germinate and grow into a prothallus with male and female structures.

The young plant grows up from the prothallus.

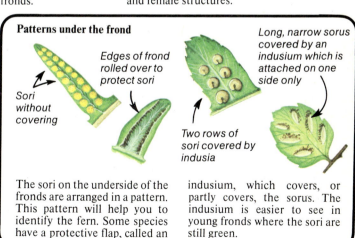

Patterns under the frond

Sori without covering
Edges of frond rolled over to protect sori
Two rows of sori covered by indusia
Long, narrow sorus covered by an indusium which is attached on one side only

The sori on the underside of the fronds are arranged in a pattern. This pattern will help you to identify the fern. Some species have a protective flap, called an indusium, which covers, or partly covers, the sorus. The indusium is easier to see in young fronds where the sori are still green.

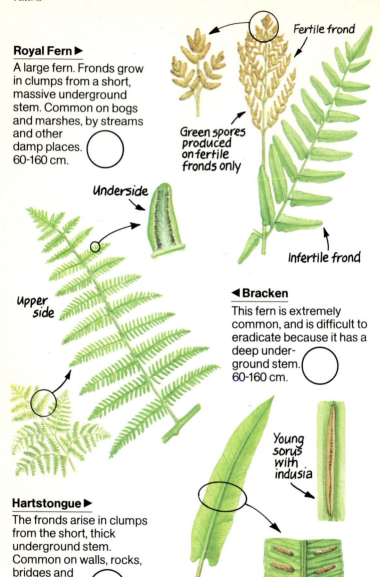

Royal Fern ▶

A large fern. Fronds grow in clumps from a short, massive underground stem. Common on bogs and marshes, by streams and other damp places. 60-160 cm.

Fertile frond

Green spores produced on fertile fronds only

Infertile frond

Underside

Upper side

◀ Bracken

This fern is extremely common, and is difficult to eradicate because it has a deep under-ground stem. 60-160 cm.

Young sorus with indusia

Hartstongue ▶

The fronds arise in clumps from the short, thick underground stem. Common on walls, rocks, bridges and in hedgerows. 10-60 cm.

Older sori

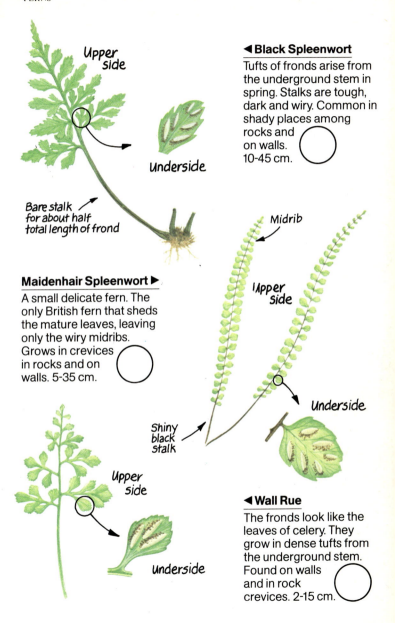

Upper side

Underside

Bare stalk for about half total length of frond

◄ Black Spleenwort
Tufts of fronds arise from the underground stem in spring. Stalks are tough, dark and wiry. Common in shady places among rocks and on walls. 10-45 cm.

Maidenhair Spleenwort ►
A small delicate fern. The only British fern that sheds the mature leaves, leaving only the wiry midribs. Grows in crevices in rocks and on walls. 5-35 cm.

Midrib

Upper side

Shiny black stalk

Underside

Upper side

Underside

◄ Wall Rue
The fronds look like the leaves of celery. They grow in dense tufts from the underground stem. Found on walls and in rock crevices. 2-15 cm.

Male Fern ▶

The commonest and most widespread British fern. Grows in a great variety of habitats, from rocky outcrops on mountains to the tops of walls in towns. 40-90 cm.

Upper side

Kidney-shaped indusium

Underside

Upper side

"Comma"-shaped indusium

Underside

Stalk is flattened, almost black at base

◀ Lady Fern

Grows in dense clumps. One of the commonest British ferns in all but the driest regions. Found mostly in damp places. 40-90 cm.

Broad Buckler Fern ▶

The fronds appear at the beginning of spring and die away each winter. Widely distributed in woods, verges, scrubland and on rocky ledges. 30-120 cm.

Upper side

Underside

Brown scales have dark central stripe

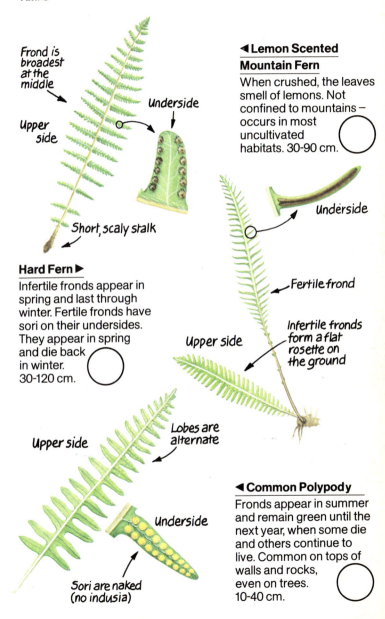

Frond is broadest at the middle

Upper side

Underside

Short, scaly stalk

Hard Fern ▶

Infertile fronds appear in spring and last through winter. Fertile fronds have sori on their undersides. They appear in spring and die back in winter. 30-120 cm.

◀ Lemon Scented Mountain Fern

When crushed, the leaves smell of lemons. Not confined to mountains – occurs in most uncultivated habitats. 30-90 cm.

Underside

Fertile frond

Infertile fronds form a flat rosette on the ground

Upper side

Upper side

Lobes are alternate

Underside

Sori are naked (no indusia)

◀ Common Polypody

Fronds appear in summer and remain green until the next year, when some die and others continue to live. Common on tops of walls and rocks, even on trees. 10-40 cm.

165

CHAPTER 8
Grasses, Rushes and Sedges

There are over 150 species of grass growing wild in the British Isles, and in this Chapter, 24 of the more common ones are described and illustrated.

The easiest way to recognize a grass is by its flowering head. This is shown in the illustrations, along with a detail of the spikelet, which is a tiny flower, or group of flowers. To find out more about the flower-heads, spikelets and flowers of grasses, study the opposite page carefully.

Grasses have long, narrow leaves with parallel veins. It is difficult to recognize a grass by its leaf alone but it can be a useful clue. Look to see if it is smooth, shiny, dull or hairy.

Grasses grow in different ways: some form tufts, others form mats spreading over the ground by means of runners or underground stems called rhizomes.

When to look
Most grasses flower in May, June and July, so these are the months when it will be easiest to spot these plants.

Where to look
Grasses form a very important part of the vegetation in many parts of the world. In Britain, you will find them on downlands, heaths, way-sides and dunes, and in woods, marshes and bogs. Don't forget that sometimes grasses do not get a chance to flower. You will not find many flowering heads on a lawn which is regularly mowed or in a field which is grazed by cattle.

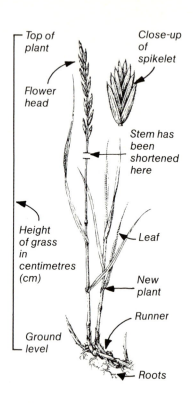

Top of plant

Close-up of spikelet

Flower head

Stem has been shortened here

Height of grass in centimetres (cm)

Leaf

New plant

Runner

Ground level

Roots

The illustrations
The illustrations of grasses in this book are in black and white but the colours of the grasses are noted around each picture. Sometimes it has been necessary to shorten the stem in order to fit it onto the page: this is shown by two short lines across the break in the stem. The height of the grass is given in the description near each illustration.

The flowering head
The flowering heads are one of the most obvious features of grasses. They vary in shape, size, colour and degree of denseness. The flowering head of a grass is made up of clusters called spikelets. The arrangement of the spikelets varies from species to species. They may be pressed close to the stalk, spaced out along it or they may hang on spreading side-branches. Grass pollen is wafted from flower to flower by the wind; this is why the flower heads are raised high above the ground on long slender stalks which wave about in the breeze.

A closer look at the flowers of grasses
A spikelet consists of a flower, or several flowers. Grass flowers are very small, pale green and do not have petals. Because grasses are pollinated by the wind, it does not matter that the flowers are inconspicuous. Most of the flowers described in Chapter 1 are pollinated by insects, so they have to be bright

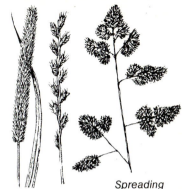

Compact head *Alternate spikelets* *Spreading side-branches*

and colourful in order to attract insects.

Some grass flowers open in the morning, others in the afternoon and some open only in the evening. If you look closely at the spikelet when the flowers are open, you will see the male and female parts (the stamens and the stigmas). A hand lens will be useful.

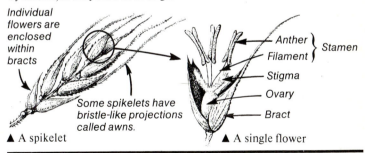

Individual flowers are enclosed within bracts

Some spikelets have bristle-like projections called awns.

▲ A spikelet

Anther ⎫ *Stamen*
Filament ⎭
Stigma
Ovary
Bract

▲ A single flower

Sedges
These are grass-like plants which usually grow by fresh water. You are likely to find them by ponds, streams and in bogs. Unlike grasses, they have solid stems which are often three-sided.

Rushes
Rushes are more closely related to lilies than they are to grasses. They grow in damp places and by water. They have small heads of greenish or brownish flowers. Their flower stalks are often tall and leafless.

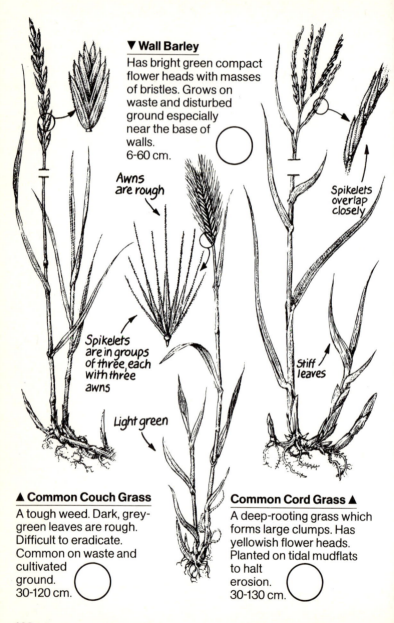

▼ Wall Barley
Has bright green compact flower heads with masses of bristles. Grows on waste and disturbed ground especially near the base of walls. 6-60 cm.

Awns are rough

Spikelets are in groups of three, each with three awns

Light green

Spikelets overlap closely

Stiff leaves

▲ Common Couch Grass
A tough weed. Dark, grey-green leaves are rough. Difficult to eradicate. Common on waste and cultivated ground. 30-120 cm.

Common Cord Grass ▲
A deep-rooting grass which forms large clumps. Has yellowish flower heads. Planted on tidal mudflats to halt erosion. 30-130 cm.

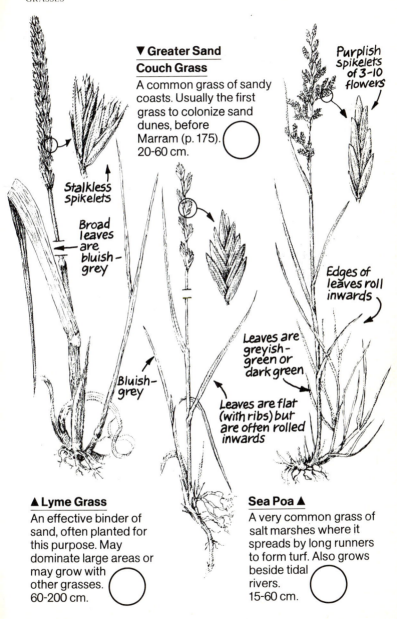

▼ Greater Sand Couch Grass

A common grass of sandy coasts. Usually the first grass to colonize sand dunes, before Marram (p. 175). 20-60 cm.

Stalkless Spikelets

Broad leaves are bluish-grey

Purplish spikelets of 3-10 flowers

Bluish-grey

Edges of leaves roll inwards

Leaves are greyish-green or dark green

Leaves are flat (with ribs) but are often rolled inwards

▲ Lyme Grass

An effective binder of sand, often planted for this purpose. May dominate large areas or may grow with other grasses. 60-200 cm.

Sea Poa ▲

A very common grass of salt marshes where it spreads by long runners to form turf. Also grows beside tidal rivers. 15-60 cm.

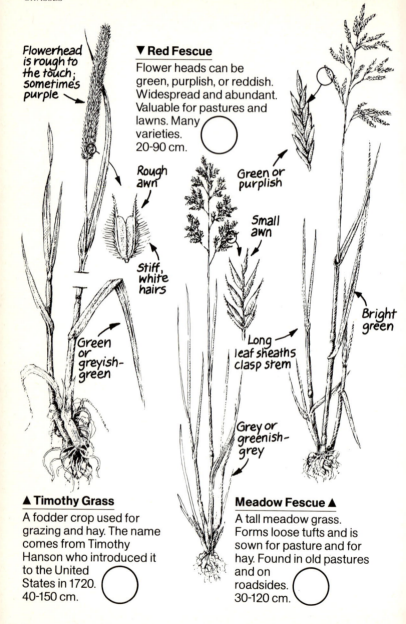

Flowerhead is rough to the touch; sometimes purple

Rough awn

Stiff, white hairs

Green or greyish-green

▼ Red Fescue
Flower heads can be green, purplish, or reddish. Widespread and abundant. Valuable for pastures and lawns. Many varieties. 20-90 cm.

Green or purplish

Small awn

Long leaf sheaths clasp stem

Bright green

Grey or greenish-grey

▲ Timothy Grass
A fodder crop used for grazing and hay. The name comes from Timothy Hanson who introduced it to the United States in 1720. 40-150 cm.

Meadow Fescue ▲
A tall meadow grass. Forms loose tufts and is sown for pasture and for hay. Found in old pastures and on roadsides. 30-120 cm.

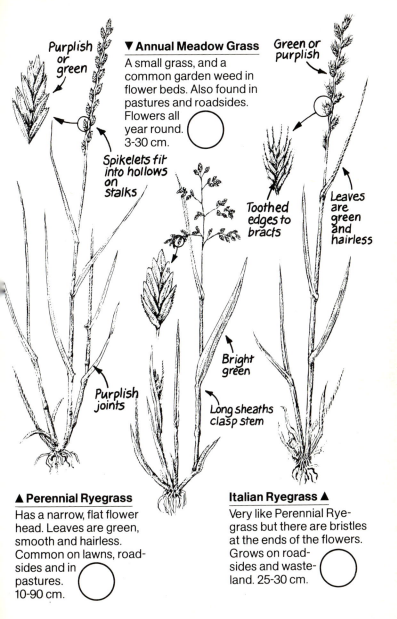

Purplish or green

▼ **Annual Meadow Grass**
A small grass, and a common garden weed in flower beds. Also found in pastures and roadsides. Flowers all year round. 3-30 cm.

Green or purplish

Spikelets fit into hollows on stalks

Toothed edges to bracts

Leaves are green and hairless

Bright green

Long sheaths clasp stem

Purplish joints

▲ **Perennial Ryegrass**
Has a narrow, flat flower head. Leaves are green, smooth and hairless. Common on lawns, roadsides and in pastures. 10-90 cm.

Italian Ryegrass ▲
Very like Perennial Ryegrass but there are bristles at the ends of the flowers. Grows on roadsides and wasteland. 25-30 cm.

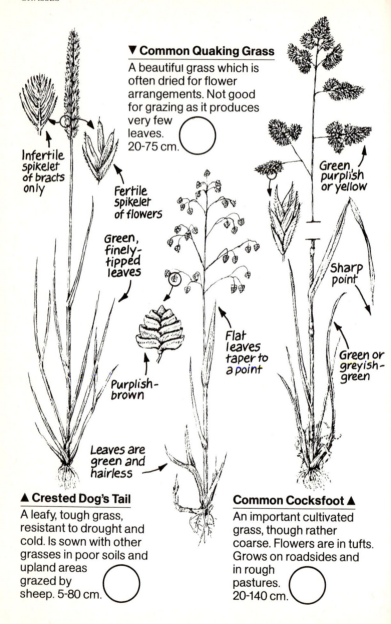

▼ Common Quaking Grass

A beautiful grass which is often dried for flower arrangements. Not good for grazing as it produces very few leaves. 20-75 cm.

Infertile spikelet of bracts only

Fertile spikelet of flowers

Green, finely-tipped leaves

Purplish-brown

Green, purplish or yellow

Sharp point

Green or greyish-green

Flat leaves taper to a point

Leaves are green and hairless

▲ Crested Dog's Tail

A leafy, tough grass, resistant to drought and cold. Is sown with other grasses in poor soils and upland areas grazed by sheep. 5-80 cm.

Common Cocksfoot ▲

An important cultivated grass, though rather coarse. Flowers are in tufts. Grows on roadsides and in rough pastures. 20-140 cm.

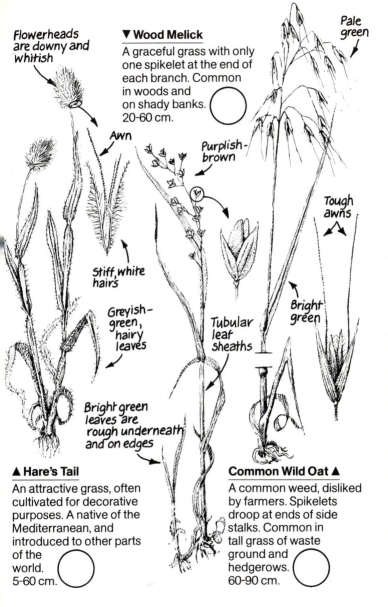

Flowerheads are downy and whitish

Awn

▼ Wood Melick
A graceful grass with only one spikelet at the end of each branch. Common in woods and on shady banks. 20-60 cm.

Purplish-brown

Pale green

Stiff, white hairs

Greyish-green, hairy leaves

Tubular leaf sheaths

Tough awns

Bright green

Bright green leaves are rough underneath and on edges

▲ Hare's Tail
An attractive grass, often cultivated for decorative purposes. A native of the Mediterranean, and introduced to other parts of the world. 5-60 cm.

Common Wild Oat ▲
A common weed, disliked by farmers. Spikelets droop at ends of side stalks. Common in tall grass of waste ground and hedgerows. 60-90 cm.

173

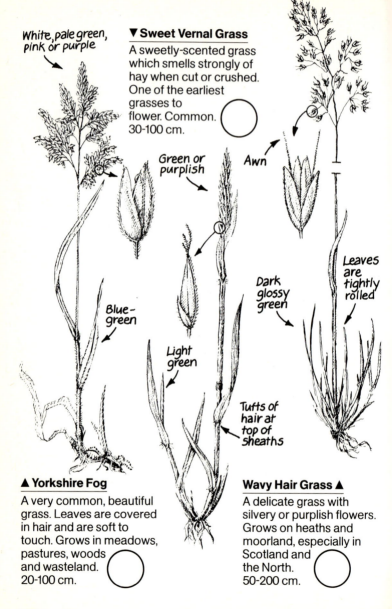

White, pale green, pink or purple

▼ Sweet Vernal Grass

A sweetly-scented grass which smells strongly of hay when cut or crushed. One of the earliest grasses to flower. Common. 30-100 cm.

Green or purplish

Awn

Blue-green

Dark glossy green

Light green

Leaves are tightly rolled

Tufts of hair at top of sheaths

▲ Yorkshire Fog

A very common, beautiful grass. Leaves are covered in hair and are soft to touch. Grows in meadows, pastures, woods and wasteland. 20-100 cm.

Wavy Hair Grass ▲

A delicate grass with silvery or purplish flowers. Grows on heaths and moorland, especially in Scotland and the North. 50-200 cm.

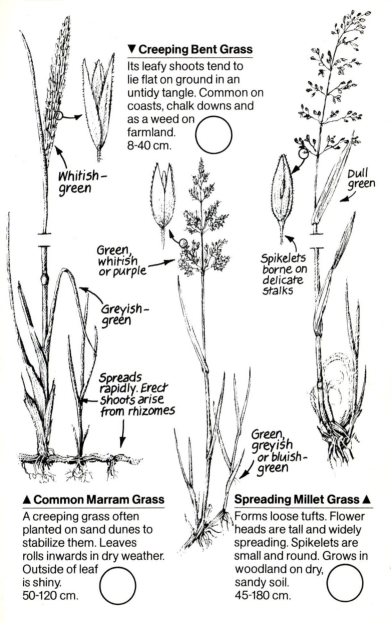

▼ Creeping Bent Grass

Its leafy shoots tend to lie flat on ground in an untidy tangle. Common on coasts, chalk downs and as a weed on farmland. 8-40 cm.

Whitish-green

Green, whitish or purple

Greyish-green

Spreads rapidly. Erect shoots arise from rhizomes

Dull green

Spikelets borne on delicate stalks

Green, greyish or bluish-green

▲ Common Marram Grass

A creeping grass often planted on sand dunes to stabilize them. Leaves rolls inwards in dry weather. Outside of leaf is shiny. 50-120 cm.

Spreading Millet Grass ▲

Forms loose tufts. Flower heads are tall and widely spreading. Spikelets are small and round. Grows in woodland on dry, sandy soil. 45-180 cm.

175

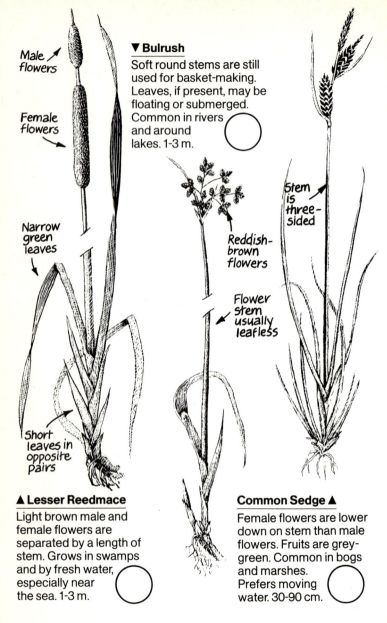

Male
flowers

Female
flowers

Narrow
green
leaves

Short
leaves in
opposite
pairs

▼ Bulrush
Soft round stems are still
used for basket-making.
Leaves, if present, may be
floating or submerged.
Common in rivers
and around
lakes. 1-3 m.

Reddish-
brown
flowers

Flower
stem
usually
leafless

Stem
is three-
sided

▲ Lesser Reedmace
Light brown male and
female flowers are
separated by a length of
stem. Grows in swamps
and by fresh water,
especially near
the sea. 1-3 m.

Common Sedge ▲
Female flowers are lower
down on stem than male
flowers. Fruits are grey-
green. Common in bogs
and marshes.
Prefers moving
water. 30-90 cm.

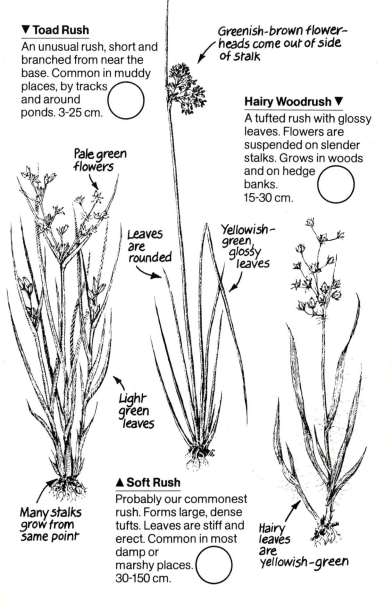

▼ Toad Rush

An unusual rush, short and branched from near the base. Common in muddy places, by tracks and around ponds. 3-25 cm.

Greenish-brown flower-heads come out of side of stalk

Hairy Woodrush ▼

A tufted rush with glossy leaves. Flowers are suspended on slender stalks. Grows in woods and on hedge banks. 15-30 cm.

Pale green flowers

Leaves are rounded

Yellowish-green glossy leaves

Light green leaves

Many stalks grow from same point

▲ Soft Rush

Probably our commonest rush. Forms large, dense tufts. Leaves are stiff and erect. Common in most damp or marshy places. 30-150 cm.

Hairy leaves are yellowish-green

Studying wild plants

Protecting wild plants

It is a sad fact that many plants once common in this country are now rare or extinct. The reasons for this are various. Over the centuries, farming practices have changed. This has been necessary in order to feed a large and increasing population, on a fairly small land area.

Fields have been enlarged, destroying hedges and woods – and therefore the wildlife that they shelter – in the process. Sprays and industrial chemicals are washed through the soil into the waterways, polluting rivers, streams and lakes. More and more land is being used for industrial estates and motorways. However, owing to the efforts of a large number of individuals and organizations, people are becoming increasingly aware of the need to protect and conserve our wildlife.

All over the country there are wildlife sanctuaries and nature reserves run by organizations like the National Trust. There may be some near you, and if you are interested in visiting some of them, your local natural history society will have details. You can obtain addresses of such societies from your local library. Not all reserves are open to the public, so it is worth checking before you set off.

How you can help

Serious destruction is caused by visitors to the countryside who pick and uproot wild plants. Through carelessness, they also cause fires which destroy large areas of common, heath and forest every year.

It is now against the law to pick certain rare plants or to dig up any wild plant by the roots. If you must pick a plant in order to study it, make sure that it is very common and pick only one specimen.

If you think you have found a rare plant, let your local County Trust for Nature Conservation know about it so that they can make arrangements for it to be protected. You can find their address at the local library or Town Hall.

The best way to record your finds is to make drawings or take photographs of them.

Go prepared

Take this book with you when you go out spotting wild plants; it is small enough to fit into a pocket or bag. You will also need a notebook in which to record your finds and a sketchbook so that you can make drawings.

Drawing a plant is the best way to examine it closely. Do a drawing to show the shape of the whole plant and the plants growing around it. Label the various parts and if you have no coloured pencils with you, make a note of the colours. Do separate drawings of interesting or unusual features. You don't have to be a great artist – anyone can produce a useful sketch – and you'll almost certainly improve with practice.

Make a list of all the things you find. A pocket ruler is useful so that you can measure the plant. A hand lens is another useful tool. It does not have to be an expensive one, and with it you can look carefully at complex structures like flowers or grass spikelets.

Photographing wild plants

Take a camera, if you have one, as photos of plants are a very good way of recording flowers. It is also important to make notes of what you are photographing, the place, the date and even the weather and the time of day.

Always use colour film, and take photos with the sun behind you; make sure your shadow does not fall on the plant. Try to get the whole plant in the picture, in the case of flowers, so that you can see the stem and leaves. Put something like a book or a pencil in the picture so that you will have a visual key to the scale of the plant.

If you think the plant you want to photograph will get lost amongst the surrounding plants and grasses, prop a piece of black or coloured card behind it as a background.

Pressing and drying

If you pick very common flowers, you can press them. Put the specimen between two sheets of blotting paper and pile some heavy books on top of it. Change the blotting paper at regular intervals to avoid the flower becoming mouldy. When it is dry-it will take several weeks-stick it into your notebook with a spot of glue. Write notes next to the specimen about the colour of the flower when fresh (it will change colour as it dries), details about the other kinds of plants growing around the specimen when you found it, and the time of year.

You could, as an alternative, keep the specimen between sheets of paper inside a box; add some mothballs to deter insects.

Collecting seaweeds

You can also collect seaweeds, since the ones you will find washed up on the beach will be dead already anyway. Do not take specimens that are still attached to their surface, and therefore still growing.

Because most seaweeds are slimy and floppy, they lose their shape out of water; the fronds mat together. It is very easy to mount seaweed onto paper. Put it in water and slide a piece of cartridge paper underneath it. The frond will have opened out and you can float it onto the paper. If you are careful, you can take the paper from the water without disturbing the seaweed, which will stick to the paper because of its slimy surface.

Cover it with a piece of muslin and several layers of blotting paper and newspaper and allow it to dry slowly under a light weight. Change the blotting paper regularly to prevent mould forming. Label the specimen and store it flat.

Bark rubbings

Bark rubbings are very useful for recording information about trees. To make one, you need a strong thin paper, wax crayons and sellotape. Tape a piece of paper against the trunk of a tree. Rub firmly up and down on the paper with the crayon until the bark pattern appears. Be careful not to tear the paper by rubbing too hard.

Glossary

Algae – Simple plants which range from minute plankton, in the sea or in freshwater, to large seaweeds.

Alternate – The arrangement of leaves on the stem in two rows which are not opposite. (See p.7.)

Annual – A plant that lives and completes its life cycle in the space of a year.

Anther - In flowers, the often brightly coloured sacs at the top of the stamen. It produces the pollen. (See p.7.)

Awn – A stiff, bristle-like projection at the tip of a flower or fruit in grasses. (See p.167.)

Bipinnate – Term used to describe a compound leaf in which the leaflets are themselves divided. (See p.7.)

Bog – An area of land that is very wet, owing to poor drainage. Some bogs are acid (e.g. peat or blanket bogs).

Bract – A leaf-like structure which grows just beneath a flower or group of flowers in some plants. (See p.26.)

Broadleaved trees – Trees that normally have broad, flat leaves and seeds which are enclosed in fruits (nuts, fleshy fruits or other forms). Most European broadleaved trees are deciduous.

Bulb – An underground structure composed of a very short stem with fleshy, food-storing leaves around it, which enclose a bud. The bud uses this food store to grow into a new plant the next year.

Calyx – The ring of sepals surrounding a flower which enclose and protect it until it opens up.

Capsule (in flowering plants) – A dry fruit which contains one or more seeds. (See p.6.)

Capsule (in mosses) – The structure which contains the spores, usually carried on a slender stalk called a seta.

Chalk grasslands – The open, treeless chalk hills of southern England, usually covered with short, compact grasses and a wide variety of other flowering plants.

Chlorophyll – The substance which gives almost all plants their green colour and enables them to manufacture food.

Compound leaf – A leaf which is made up of several smaller leaflets. (See p.7 and see also Simple leaf.)

Cone – The woody "fruit" of a coniferous tree, formed of overlapping scales.

Conifers – Trees which usually have narrow, needle-like leaves, or flattened, scaly leaves which lie close to the stem. Most conifers shed their leaves a few at a time, all the year round, so the tree is never bare. Their seeds are produced in cones.

Corolla – The brightly coloured part of a flower made up of the petals. The petals may be separate, or fused together forming, for example, a tube.

Crown – The leafy branch system of a tree.

Crozier – The young, curled-up frond of a fern plant, so called because it resembles a bishop's crozier. (See p.160.)

Cushion – The dense, neat mound formed by some mosses.

Deciduous – Term which is used to describe trees which drop all their leaves in the autumn.

Ecology – The study of the relationship between plants and animals and their environment.

Egg – The female reproductive cell in lower plants, such as mosses and ferns.

Entire – Term used to describe the edges of leaves which are not lobed or toothed.

Erect – Term used to describe a plant which grows straight up from the ground.

Erosion – The process by which soil or rock is broken down by heat, wind and water.

Evergreen – Term used to describe trees that do not shed all their leaves in autumn. (See Deciduous.)

Fertilization – In sexual reproduction, the process by which a female cell (egg) is united with a male cell (sperm). The resulting cell eventually forms a new plant.

Fiddlehead – See Crozier.

Filament (of lower plants) – A string of single cells produced by germinating spores.

Filament (in stamens) – The stalk of the stamen. It carries the anther. (See p.7.)

Flowering head – In flowering plants, groups or clusters of flowers. In some cases, the flowers are so tightly packed that the whole thing looks like a single flower, e.g. Dandelion. Another term for the flowering head is the "inflorescence".

Frond – The leaf-like structure in plants such as seaweeds and ferns (See pp. 146, 160.)

Fruit – The ripened ovary of a flowering plant. It contains the seed or seeds. It may be fleshy or dry. (See also Capsule.)

Fruiting body – The spore-producing structure of lichens, mosses, liverworts and fungi. It is the part of some fungi which most people recognize as being a mushroom or toadstool.

Fungus (plural: fungi) – A large group of flowerless "plants". Most people regard them as members of the plant kingdom, despite the fact that they cannot manufacture their own food in the way that most green plants can.

Gemmae – The structures often found in little cups called gemmae cups in mosses and liverworts. The gemmae can grow into new plants. They are different from the spores, which are also found in mosses and liverworts.

Germination – The initial growth of a seed or spore to become a new plant. In seeds this is seen as the development of first the roots and then the shoot.

Gills – The ribs on the underside of some fungi which radiate out from the stem like the spokes of a wheel.

Habitat – The situation in which a plant or animal lives. It includes all other organisms, the soil and the climate.

Heath – Sandy, gravelly areas of acid soil. Often has heather growing on it.

Holdfast – In seaweeds the disc- or root-like structure by which the plant is attached to the surface on which it grows. (See p.147.)

Hyphae – The fine threads which make up a fungal mycelium and form the largest part of the thallus of a lichen.

Indusium (plurial: indusia) – In ferns the protective covering of the sorus. (See p.161.)

Intermediate gill – A gill that does not run all the way from the edge of a mushroom cap to the stem.

Lamina – The blade of a leaf.
Leaflet – Small leaf that forms part of a compound leaf. (See p.7.)
Leafy liverworts – Liverworts with creeping (not upright) stems and leaves. (See p.153 and see also Thallose liverworts.)
Lobe – A segment or division of, for example, a leaf, but not one that is separate from other parts of the leaf, as in leaflets. (See pp. 7, 49.)
Local – Restricted to certain areas. Not widespread.
Lower shore – The area between the low water lines of the spring and neap tides. (See p. 147.)

Marsh – Waterlogged ground. It can be salt or fresh water.
Mat-forming – Term used for plants which grow close to the ground forming thick mats or carpets.
Middle shore – The main part of the beach between the lower shore and the upper shore. (See p. 147.)
Midrib – The central vein of a leaf.
Mushroom – A cap fungus with gills that belongs to the *Agaricus* group. Some people call all edible cap fungi "mushrooms".
Mycelium – The non-reproductive part of a fungus consisting of a mass of fine interwoven threads called hyphae. It lives in the soil or in the plant or animal matter upon which it feeds. Fruiting bodies grow up from the mycelium – these are the parts of the fungus we see above ground.

Neap tides – Very small tides which occur, at fortnightly intervals, between spring tides.
Nutrients – The substances in soil upon which plants feed.

Opposite – The arrangement of leaves growing in opposite pairs on the stem. (See p.7.)
Ovary – The part of the female reproductive structure of the flower which contains the egg cell and later the seeds. (See p.7.)

Palmate – Term used to describe a compound leaf in which the leaflets are arranged like the fingers on a hand. (See p.7.)
Peat – Soil formed from plant matter under wet conditions. Cut and dried, it can be used for fuel.
Perennial – A plant that lives for more than two years, usually flowering each year.
Petals – The individual parts of the corolla of a flower – usually brightly coloured. (See p.7.)
Pinnate – Term used to describe a compound leaf in which the two rows of leaflets are arranged in pairs on either side of the midrib. The leaf can look feather-like. (See p.7.)
Pollen – The yellow powder which contains the male reproductive cells of flowering plants and of conifers. It is produced in the anthers of flowering plants and of special structures, sometimes called "flowers", in conifers.
Pollination – The process by which pollen is carried by wind or insects from the anthers to the stigmas of flowers.
Pores – The round openings of tubes, on the underside of the cap of *Boletus* species and of most bracket fungi.
Prothallus – In ferns, the tiny, green, heart-shaped structure which bears the male and female reproductive organs. (See p.161.)

Rhizines – The fine threads by which a lichen is attached to the surface on which it grows.

Rhizoids – The fine, root-like threads by which a moss is attached to the surface on which it grows. (See p.152.)

Rhizome – An underground stem which can become swollen with stored food for the plant to use for new growth each year.

Ring (in fungi) – The remains of the veil left on the stalk of a cap fungus.

Rosette – A ring of leaves, usually growing very close to the ground surface; produced by plants with very short stems.

Runner – A stem which creeps over the surface of the ground, rooting at intervals and producing a new plant.

Seed – The part of a flowering plant which can grow into a new plant. It is usually contained within a nut or fruit.

Sepals – The individual units which make up the calyx of a flower. (See p.7.)

Simple leaf – A leaf which is undivided.

Sorus (plural: sori) – In ferns, a group of sporangia, the organs which contain the spores. (See p.161.)

Sperm cell – Male reproductive cell.

Spikelet – The flower unit of a grass. It may consist of one or several flowers. Usually green and inconspicuous, not brightly coloured like the flowers of most other plants, as it has no petals. (See p.167.)

Spine (in flowering plants) – a woody, thorn-like projection.

Spines (in fungi) – Small, tooth-like projections which occur under the cap of some fungi instead of gills.

Spores – Minute reproductive cells of plants which do not produce seeds, such as mosses, ferns, fungi and lichens.

Spring tides – Very high and very low tides which occur at fortnightly intervals.

Spur – A round hollow projection formed by the fusion of flower petals in some species, e.g. Violet.

Stamen – The male part of a flower. It consists of the stalk, called a filament, and the anther.

Stigma – The tip of the female reproductive part of a flower. It is where the pollen is trapped.

Stipe – The stalk of a seaweed frond. (See p.146.)

Symbiosis – Two organisms living together and both partners benefitting from the relationship.

Teeth (in mosses) – Specialized parts of the capsule known as peristome teeth which can project the spores but can also open to release the spores when conditions are right for spore dispersal.

Thallose liverworts – Liverworts with no distinct stems or leaves. (See p.153.)

Thallus – The body of simple plants such as lichens, seaweeds and some liverworts. It is not divided into separate organs such as stems and leaves.

Toadstool – A general term for any cap fungus with a stem.

Toothed – Term used to describe the edge of a leaf when it is jagged.

Tuft – A cluster of shoots growing from the same point.

Upper shore – The area between the high water lines of the spring and neap tides. (See p.147.)

Veil – The outer "skin" on the underside of the cap of a young toadstool that splits to reveal the gills.

Volva – In fungi, the cup-like remains of the veil at the base of the stem.

Whorl – A ring of three or more flowers or leaves around a stem. (See p.7.)

Index